Artin.
Introduction to algebraic
topology.

Date Due

INTRODUCTION TO ALGEBRAIC TOPOLOGY

MERRILL RESEARCH AND LECTURE SERIES

Erwin Kleinfeld, *Editor*

INTRODUCTION TO ALGEBRAIC TOPOLOGY

EMIL ARTIN and HEL BRAUN
University of Hamburg

Translated from the notes of
Armin Thedy and Hel Braun
by

ERIK HEMMINGSEN
Syracuse University

CHARLES E. MERRILL PUBLISHING COMPANY
A Bell & Howell Company
Columbus, Ohio

Library of Congress Catalog Card Number: 69–11166
Standard Book Number: 675–096510–0
AMS Classification Number: 5501

PRINTED IN THE UNITED STATES OF AMERICA

1 2 3 4 5 6 7 8 9 10—73 72 71 70 69

Foreword to the German Edition

Professor Emil Artin lectured on algebraic topology during the winter semester of 1959–60. Later, in the summer of 1962 and the winter semester of 1962–63, Professor Hel Braun gave a more extensive course on the same subject in which she made her preparation in consultation with Professor Artin. The present volume is based on lecture notes of Professor Hel Braun, which she very kindly placed at my disposal. I wish to express my particular gratitude for this.

Armin Thedy

Foreword to the English Edition

The first nineteen sections of this volume are a revised translation of the German edition. The last three sections are new and are taken from the notes of Hel Braun. Basic changes have been avoided in the hope of preserving, as much as possible, the flavor of Artin's original presentation.

The reader is expected to have some familiarity with the elements of point set topology and an introduction to the theory of groups, rings, and vector spaces. However, the book is remarkably self-contained. Proofs are given of many of the algebraic and topological theorems that are central to the subjects discussed.

The early sections introduce exact sequences of modules, the algebraic structure of homology groups of chain complexes, and the geometry of affine spaces and simplices. The authors then proceed directly to the discussion of singular homology, for which the theorems corresponding to the Eilenberg-Steenrod axioms are proved. The middle third of the book is devoted to tensor products, categories, functors, axiomatic homology theory, and the Mayer-Vietoris Theorem. The last third contains a number of geometrical applications (the Jordan-Brouwer separation theorem, finite cell complexes, betti numbers and euler characteristics, projective spaces, degree of mappings of spheres, lens spaces, and the classification of compact separable two-manifolds) together with an introduction to cup and cap products. However, not all applications occur at the end of the book. From the very early pages, there are extensive geometrical applications which serve to give the reader a good intuitive grasp of the meaning and effectiveness of the tools that have been developed.

Erik Hemmingsen

Table of Contents

Chapter 1
Homology Groups of Chain Complexes 1

Chapter 2
Affine Spaces 10

Chapter 3
Affine Simplices and Boundary Operator 17

Chapter 4
The Singular Homology Theory 27

Chapter 5
Homotopy Properties of Homology Groups 38

Geometrical Consequences of the Homotopy Theorem 50
Application to Graphs 53

Chapter 6
The Excision Theorem 57

Chapter 7
Direct Decomposition and Additional Aids to the Computation of Homology Groups 70

Computation of $H_0(X, A)$ 72
Homology Groups of a Point P 73
The Sphere 75

Homology Groups of Graphs 80
Degree of a Function $f:S^n \to S^n$ 81

Chapter 8
The Tensor Product **85**

Tensor Products of Functions 91

Chapter 9
The Functor Hom **98**

R, S—Modules 101
Quotient Modules 103

Chapter 10
Categories and Functors **107**

Chapter 11
Categories, Functors, and the Singular Theory **115**

The Functors ⊗ *M* and Hom (,*M*) in the Singular Theory 118
Homology 119
Cohomology 120
Application of Homology to Function Theory 122

Chapter 12
Axioms for Homology and Cohomology **129**

Chapter 13
Mayer-Vietoris Sequence **137**

Preliminary Considerations with Respect to Modules 137

Chapter 14
The Jordan-Brouwer Separation Theorems **149**

Chapter 15
Finite Cell Complexes 158

Spherical Complexes 164
Products of Spherical Complexes 166

Chapter 16
Betti Numbers and the Euler Characteristics 169

Chapter 17
Complex and Real Projective Spaces 175

Groups of the Complex Projective Space CP^n 177
Groups of the Real Projective Space P^n 178

Chapter 18
Maps of S^n on S^n and Lens Spaces 182
Application to Lens Spaces 187

Chapter 19
Classification of Surfaces 193

Simplicial Complexes 194

Chapter 20
Singular Cup Products 201

Chapter 21
The Singular Cap Product 209

Chapter 22
The Anticommutativity of the Cup Product 216

Index 225

1

Homology Groups of Chain Complexes

Let R be a commutative and associative ring with unit element 1. By a (unitary) *R-module* is meant an additively written, abelian group A, for which a product $R \times A \longrightarrow A$ is defined such that

$$r(a + a') = ra + ra', \qquad (r + r')a = ra + r'a \qquad r, r' \in R$$
$$r(r'a) = (rr')a, \qquad 1a = a \qquad a, a' \in A.$$

Let A, B be R-modules. A function $f:A \longrightarrow B$ of A into B is called an R-homomorphism (or merely a homomorphism, if it is clear which R is under discussion) if and only if $f(a + a') = f(a) + f(a')$ and $f(ra) = rf(a)$ hold for $a, a' \in A$, and $r \in R$. The set of all $a \in A$ with $f(a) = 0$ is called the *kernel of f* and designated by $\text{Ker}(f) = f^{-1}(0)$.

The set $f(A)$ is designated by Im(f). The homomorphism $f:A \longrightarrow B$ is called an *epimorphism* if and only if $f(A) = B$. It is called a *monomorphism* if and only if $\operatorname{Ker} f = 0$, and it is called an *isomorphism* if and only if it is both an epimorphism and a monomorphism. The kernel of $f:A \longrightarrow B$ is a submodule of A. If C is a submodule of A, then A/C is made into an R-module, the factor module of A mod C, by defining the sum and product operations in terms of representatives:

$$(a + C) + (a' + C) = (a + a')C, \quad r(a + C) = ra + C.$$

To each $f:A \longrightarrow B$ corresponds the canonical decomposition $f = f_2 f_1 f_0$ where

I. $f_0:A \longrightarrow A/f^{-1}(0)$ is defined by $f_0(a) = a + f^{-1}(0)$ and is an epimorphism,

II. $f_1:A/f^{-1}(0) \longrightarrow f(A)$ is defined by $f_1(a + f^{-1}(0)) = f(a)$ and is an isomorphism,

III. $f_2:f(A) \longrightarrow B$ is defined by $f_2(f(a)) = f(a)$ and is a monomorphism.

It is easy to prove:

1.1. Lemma: *Let $f:A \longrightarrow B$ be a homomorphism of R-modules, C a submodule of A, and D a submodule of B. Then f induces a homomorphism, $f:A/C \longrightarrow B/D$ with $f(a + C) = f(a) + D$ if and only if $f(C) \subset D$.*

The set of all homomorphisms of A in B is denoted by $\operatorname{Hom}_R(A, B)$ or by Hom(A, B). Hom(A, B) becomes an R-module when the definitions

$$(f + f')(a) = f(a) + f'(a), \quad (rf)a = rf(a)$$

are made for all $f, f' \in \operatorname{Hom}(A, B)$, $a \in A$, and $r \in R$.

Let Z be the set of integers. A collection of R-modules A_q and R-homomorphisms f_q, $q \in Z$ such that

$$\cdots \longrightarrow A_{q+1} \xrightarrow{f_{q+1}} A_q \xrightarrow{f_q} A_{q-1} \longrightarrow \cdots$$

is called a *sequence*. Here, q can run through all the integers, or else the sequence can terminate on the right, on the left, or on both sides. It is also possible for the arrows all to be directed from right to left.

A sequence is called *exact at the location q* if and only if $\text{Im}(f_{q+1}) = f_{q+1}(A_{q+1}) = f_q^{-1}(0) = \text{Ker}(f_q)$. A sequence is called *exact* if and only if it is exact at all locations having one arrow leading in and one arrow leading out. Exactness at the location A_q is illustrated schematically by Figure 1.

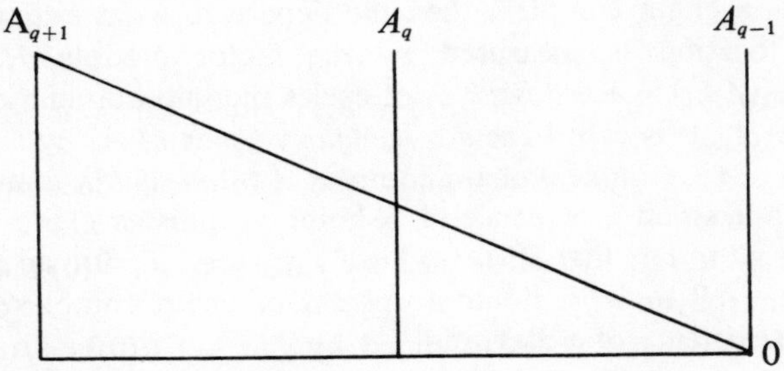

Figure 1

To prove exactness at the location A_q, the following must be established:

(1) $f_{q+1}(A_{q+1}) \subset \text{Ker}(f_q)$; that is, $f_q f_{q+1} = 0$; and
(2) $f_{q+1}(A_{q+1}) \subset \text{Ker}(f_q)$; that is, to each $a_q \in A_q$, for which $f_q(a_q) = 0$, there is an $a_{q+1} \in A_{q+1}$ for which $f_{q+1}(a_{q+1}) = a_q$.

The module consisting of the zero element only is designated by 0. By $0 \longrightarrow A$ is to be understood that injection of the zero module into A for which the image is the zero element of A. Furthermore, $B \longrightarrow 0$ designates the function under which all elements of B have as image the element 0. The following special cases of exactness are important:

(1) $0 \longrightarrow A \xrightarrow{i} B$ exact means exactness at A; that is, that $\text{Ker}(i) = 0$ and that i is a monomorphism.

(2) $A \xrightarrow{j} B \longrightarrow 0$ exact means exactness at B; that is, that $j(A) = B$ and that j is an epimorphism.

(3) $0 \longrightarrow A \longrightarrow 0$ exact means $A = 0$.

(4) $0 \longrightarrow A \xrightarrow{i} B \longrightarrow 0$ exact means that i is an isomorphism.

(5) $0 \longrightarrow A \xrightarrow{i} B \xrightarrow{j} C \longrightarrow 0$ exact means that i is a monomorphism, j is an epimorphism, and that $\text{Im}(i) = \text{Ker } j$. Thus, C is isomorphic to $B/\text{Im}(i)$.

A sequence $\cdots \longrightarrow A_{q+1} \xrightarrow{\partial_{q+1}} A_q \xrightarrow{\partial_q} A_{q-1} \longrightarrow \cdots$ in which $\partial_q\partial_{q+1} = 0$ for all q is called a *chain complex*. The elements of $\partial_q^{-1}(0) = \text{Ker}(\partial_q)$ are called *cycles*, the elements of $\partial_{q+1}(A_{q+1}) = \text{Im}\partial_{q+1}$ are called *boundaries*, and $\partial_{q+1}(A_{q+1}) \subset \text{Ker}(\partial_q)$. In most cases the subscript q will be omitted from ∂_q, and the symbol ∂ will be called a *boundary operator*.

If A is a chain complex, then the departure from exactness at the qth location is measured by the factor module $H_q(A) = \text{Ker}\,\partial_q/\partial_{q+1}(A_{q+1}) = \text{Ker}\partial_q/\text{Im}\partial_{q+1}$ of cycles modulo boundaries. The R-module $H_q(A)$ is called the qth *homology group of A*. By a *homomorphism* $g:A \longrightarrow B$ of a chain complex A into a chain complex B is to be understood a sequence of R-homomorphisms $g_q:A_q \longrightarrow B_q$ for which $g\partial = \partial g$; that is, for which $\partial_q g_q = g_{q-1}\partial_q$ for all $q \in Z$. For $g:A \longrightarrow B$ to be a homomorphism of chain complexes thus means the existence of a diagram

$$\begin{array}{ccccccccc} \cdots & \longrightarrow & A_{q+1} & \longrightarrow & A_q & \longrightarrow & A_{q-1} & \longrightarrow & \cdots \\ & & \downarrow{\scriptstyle g_{q+1}} & & \downarrow{\scriptstyle g_q} & & \downarrow{\scriptstyle g_{q-1}} & & \\ \cdots & \longrightarrow & B_{q+1} & \longrightarrow & B_q & \longrightarrow & B_{q-1} & \longrightarrow & \cdots, \end{array}$$

in which the condition, $\partial_q g_q = g_{q-1}\partial_q$ for all q, is described in words by saying that the diagram is *commutative* at each of its rectangles.

Each homomorphism $g:A \longrightarrow B$ of chain complexes induces homomorphisms $g_{q*}: H_q(A) \longrightarrow H_q(B)$ which are defined as follows: For $a_q + \partial(A_{q+1}) \in H_q(A)$,

$$g_{q*}(a_q + \partial(A_{q+1})) = g_q(a_q) + \partial(B_{q+1}).$$

The right-hand side is an element of $H_q(B)$, since

$$\partial g_q(a_q) = g_{q-1}\partial a_q = g_{q-1}(0) = 0.$$

By Lemma 1.1, the function g_{q*} is well defined since

$$g_{q*}\partial(A_{q+1}) = \partial_{q+1}(A_{q+1}) \subset \partial B_{q+1}.$$

The index q is usually omitted from g_{q*} to yield the notation

$$g_*:H_q(A) \longrightarrow H_q(B).$$

Consider the diagram $0 \longrightarrow A \xrightarrow{i} B \xrightarrow{j} C \longrightarrow 0$ of chain complexes and of chain homomorphisms i, j. Written in detail, this becomes

$$
\begin{array}{ccccccccc}
\vdots & & \vdots & & \vdots & & \vdots & & \vdots \\
\downarrow & & \downarrow & & \downarrow & & \downarrow & & \downarrow \\
0 & \longrightarrow & A_{q+1} & \xrightarrow{i_{q+1}} & B_{q+1} & \xrightarrow{j_{q+1}} & C_{q+1} & \longrightarrow & 0 \\
\partial\downarrow & & \partial\downarrow & & \partial\downarrow & & \partial\downarrow & & \partial\downarrow \\
0 & \longrightarrow & A_{q} & \xrightarrow{i_{q}} & B_{q} & \xrightarrow{j_{q}} & C_{q} & \longrightarrow & 0 \\
\partial\downarrow & & \partial\downarrow & & \partial\downarrow & & \partial\downarrow & & \partial\downarrow \\
0 & \longrightarrow & A_{q-1} & \xrightarrow{i_{q-1}} & B_{q-1} & \xrightarrow{j_{q-1}} & C_{q-1} & \longrightarrow & 0 \\
\downarrow & & \downarrow & & \downarrow & & \downarrow & & \downarrow \\
\vdots & & \vdots & & \vdots & & \vdots & & \vdots
\end{array},
$$

where the diagram commutes at each square and ∂ is a boundary operator; that is, $\partial\partial = 0$.

The sequence $0 \longrightarrow A \xrightarrow{i} B \xrightarrow{j} C \longrightarrow 0$ of chain complexes is called exact if and only if each of the sequences $0 \longrightarrow A_q \xrightarrow{i_q} B_q \xrightarrow{j_q} C_q \longrightarrow 0$ is exact.

The homomorphisms $i_*: H_q(A) \longrightarrow H_q(B)$ and $j_*: H_q(B) \longrightarrow H_q(C)$ that correspond to i and j have already been defined. Now let $0 \longrightarrow A \longrightarrow B \longrightarrow C \longrightarrow 0$ be an exact sequence. Then a set of homeomorphisms $\partial_{q*}: H_q(C) \longrightarrow H_{q-1}(A)$ is defined as follows: The elements of $H_q(C)$ have the form $c_q + \partial C_{q+1}$, where $\partial c_q = 0$. Since j_q is an epimorphism, there is a $b_q \in B_q$ for which $j_q(b_q) = c_q$. From now on, i will be written instead of i_q and j instead of j_q. On account of $j(\partial b_q) = \partial(jb_q) = \partial c_q = 0$ and the exactness at the qth location, there is an a_{q-1} in A_{q-1} for which $ia_{q-1} = \partial b_q$.

$$
\begin{array}{ccc}
 & & b_q \xrightarrow{j} c_q \\
 & & \downarrow \\
a_{q-1} & \longrightarrow & \partial b_q
\end{array}
$$

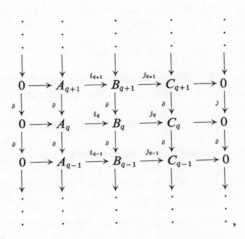

The element $\partial_{q*}(c_q + \partial C_{q+1})$ is defined to be $a_{q-1} + \partial A_q$. This is in $H_{q-1}(A)$ because $i\partial a_{q-1} = \partial i a_{q-1} = \partial\partial b_q = 0$ and thus, since i is a monomorphism, $\partial a_{q-1} = 0$. The symbol ∂_* will be written in place of ∂_{q*}.

Now it must be proved that ∂_* is well defined: Let $c'_q = c_q + \partial c_{q+1}$. Then there is a $b_{q+1} \in B_{q+1}$ for which $c_{q+1} = jb_{q+1}$ and

$$c'_q = c_q + \partial jb_{q+1} = j(b_q + \partial b_{q+1}).$$

Since $\operatorname{Ker} j_q = i(A_q)$, the most general $b'_q \in B_q$ for which $jb'_q = c'_q$ has the form $b'_q = b_q + \partial b_{q+1} + ia_q$ for a suitable $a_q \in A_q$. Then

$$\partial b'_q = \partial b_q + \partial ia_q = ia_{q-1} + i\partial a_q = i(a_{q-1} + \partial a_q).$$

Since i is a monomorphism, it follows that

$$\partial_*(c'_q + \partial C_{q+1}) = a_{q-1} + \partial a_q + \partial A_q = a_{q-1} + \partial A_q.$$

Therefore, ∂_* is well defined. It is trivial that $\partial_*: H_q(C) \longrightarrow H_{q-1}(A)$ is a homomorphism, since all calculations were linear.

1.2. Lemma: *Let A, B, C be chain complexes. Let $f:A \longrightarrow B$ and $g:B \longrightarrow C$ be chain homomorphisms. Then gf is also a chain homomorphism, and $(gf)_* = g_*f_*$. Let f,g: $A \longrightarrow B$ be chain homomorphisms. Then $f + g$ is a chain homomorphism, and $(f + g)_* = f_* + g_*$. Furthermore, $Id_* = Id$ and $0_* = 0$.*

Proof: Lemma 2 follows immediately from the definition of $_*$.

1.3. THEOREM: *If $0 \longrightarrow A \xrightarrow{i} B \xrightarrow{j} C \longrightarrow 0$ is an exact sequence of chain complexes, then*

$$\cdots \longrightarrow H_q(A) \xrightarrow{i_*} H_q(B) \xrightarrow{j_*} H_q(C)$$
$$\xrightarrow{\partial_*} H_{q-1}(A) \xrightarrow{i_*} H_{q-1}(B) \longrightarrow \cdots$$

is an exact sequence of R-modules.

Proof: The two conditions for exactness required in the definition will now be established.

(1) From Lemma 2, $j_*i_* = (ji)_* = 0_* = 0$.

(2) Let $b_q + \partial B_{q+1} \in H_q(B)$. Then

$$\partial b_q = 0 \text{ and } j_*(b_q + \partial B_{q+1}) = j(b_q) + \partial C_{q+1}.$$

In accordance with the definition of ∂_*, the element b_q is chosen as an inverse image of $j(b_q)$, and an element $a_{q-1} \in A_{q-1}$ is found for which $i(a_{q-1}) = \partial b_q$. Hence, $i(a_{q-1}) = 0$. Since i is a monomorphism, it follows that $a_{q-1} = 0$. Therefore,

$$\partial_* j_*(b_q + \partial B_{q+1}) = \partial_*(j(b_q) + \partial C_{q+1}) = a_{q-1} + \partial A_q = 0 + \partial A_q.$$

This means that $\partial_* j_* = 0$.

(3) Let $c_q + \partial C_{q+1} \in H_q(C)$. If $c_q = j(b_q)$ and $\partial b_q = i(a_{q-1})$, then $\partial_*(c_q + \partial C_{q+1}) = a_{q-1} + \partial A_q$. The image of this under i_* is consequently

$$i(a_{q-1}) + \partial B_q = \partial b_q + \partial B_q = \partial B_q.$$

This is the zero element of $H_{q-1}(B)$, and therefore, $i_* \partial_* = 0$. Thus, it has just been proved that the sequence is a chain complex.

(4) Let $a_q + \partial A_{q+1} \in \text{Ker}(i_*)$; that is, $\partial(a_q) = 0$. Then $i(a_q) = \partial b_{q+1}$ for some $b_{q+1} \in B_{q+1}$. Let $c_{q+1} = j(b_{q+1})$. Then,

$$\partial(c_{q+1}) = \partial j(b_{q+1}) = j\partial(b_{q+1}) = ji(a_q) = 0$$

since $ji = 0$. Therefore, $c_{q+1} + \partial C_{q+2} \in H_{q+1}(C)$ and $\partial_*(c_{q+1} + \partial C_{q+2}) = a_q + \partial A_{q+1}$.

(5) Let $b_q + \partial B_{q+1} \in \text{Ker}(j_*)$; that is, $j(b_q) + \partial C_{q+1} = 0$ and $\partial b = 0$. Then $j(b_q) = \partial c_{q+1}$ for some $c_{q+1} \in C_{q+1}$. Since j is an epimorphism, there is a b_{q+1} for which $j(b_{q+1}) = c_{q+1}$. Let $b'_q = b_q - \partial b_{q+1}$. Then,

$$j(b'_q) = j(b_q) - j\partial(b_{q+1}) = j(b_q) - \partial c_{q+1} = j(b_q) - j(b_q) = 0.$$

Consequently, there exists an a_q for which $i(a_q) = b'_q$. From

$$i\partial(a_q) = \partial i(a_q) = \partial(b'_q) = \partial(b_q) = 0$$

it follows that $\partial(a_q) = 0$, since i is a monomorphism. Hence, $a_q + \partial A_{q+1} \in H_q(A)$. Then,

$$i_*(a_q + \partial A_{q+1}) = i(a_q) + \partial B_{q+1} = b'_q + \partial B_{q+1} = b_q + \partial B_{q+1};$$

in other words, $\text{Ker}(j_*) \subset \text{Im}(i_*)$.

(6) Let $c_q + \partial C_{q+1} \in \text{Ker}(\partial_*)$; that is, $\partial_*(c_q + \partial C_{q+1}) = 0$ and $\partial c_q = 0$. From the choice of b_q and a_{q-1} such that $c_q = j(b_q)$ and $\partial(b_q) = i(a_{q-1})$ there follows

$$\partial_*(c_q + \partial C_{q+1}) = a_{q-1} + \partial A_q = 0.$$

Hence, there is an $a_q \in A_q$ for which $a_{q-1} = \partial a_q$. For $b'_q = b_q - i(a_q)$ it follows, therefore, that

$$\partial(b'_q) = \partial(b_q) - \partial i(a_q) = \partial(b_q) - i(a_{q-1}) = 0.$$

Hence, $b'_q + \partial B_{q+1} \in H_q(B)$. Furthermore,

$$\begin{aligned} j_*(b'_q + \partial B_{q+1}) &= j(b'_q) + \partial C_{q+1} \\ &= j(b_q) - ji(a_q) + \partial C_{q+1} = c_q + \partial C_{q+1}. \end{aligned}$$

This completes the proof of Theorem 1.3.

The following theorem is used in connection with Theorem 1.3.

1.4. THEOREM: *In the diagram of chain complexes*

$$\begin{array}{ccccccccc} 0 & \longrightarrow & A & \xrightarrow{i} & B & \xrightarrow{j} & C & \longrightarrow & 0 \\ & & \downarrow f & & \downarrow g & & \downarrow h & & \\ 0 & \longrightarrow & A' & \xrightarrow{i'} & B' & \xrightarrow{j'} & C' & \longrightarrow & 0 \end{array}$$

let the two rows be exact, and let each square be commutative; then in the diagram of R-modules

$$\begin{array}{ccccccccccc} \cdots \longrightarrow & H_q(A) & \xrightarrow{i_*} & H_q(B) & \xrightarrow{j_*} & H_q(C) & \xrightarrow{\partial_*} & H_{q-1}(A) & \xrightarrow{i_*} & H_{q-1}(B) & \longrightarrow \cdots \\ & \downarrow f_* & & \downarrow g_* & & \downarrow h_* & & \downarrow f_* & & \downarrow g_* & \\ \cdots \longrightarrow & H_q(A') & \xrightarrow{i'_*} & H_q(B') & \xrightarrow{j'_*} & H_q(C') & \xrightarrow{\partial_*} & H_{q-1}(A') & \xrightarrow{i'_*} & H_{q-1}(B') & \longrightarrow \cdots \end{array}$$

each square commutes.

Proof: Theorem 1.3 is not used here.

(1) From $i'f = gi$ and Lemma 1.2 follows $(i'f)_* = (gi)_* = i'_*f_* = g_*i_*$. The commutativity of the second square follows in the same way.

(2) In the completion of the proof, there remains the commutativity of

$$\begin{array}{ccc} H_q(C) & \xrightarrow{\partial_*} & H_{q-1}(A) \\ \downarrow h_* & & \downarrow f_* \\ H_q(C') & \xrightarrow{\partial^*} & H_{q-1}(A'). \end{array}$$

For this, $f_*\partial_*(c_q + \partial C_{q+1})$ must be computed. Therefore, let $c_q = j(b_q)$ and $\partial b_q = i(a_{q-1})$. Then

$$f_*\partial_*(c_q + \partial C_{q+1}) = f(a_{q-1}) + \partial A'_q.$$

On the other hand,

$$\partial_* h_*(c_q + \partial C_{q+1}) = \partial_*(h(c_q) + \partial C'_{q+1}).$$

By hypothesis,

$$h(c_q) = hj(b_q) = j'g(b_q)$$

and

$$\partial g(b_q) = g\partial(b_q) = gi(a_{q-1}) = i'f(a_{q-1}).$$

Then $g(b_q)$ can be used in setting up ∂_*, and it follows that

$$\partial_*(h(c_q) + \partial C'_{q+1}) = f(a_{q-1}) + \partial A'_q.$$

A comparison of those results yields $f_*\partial_* = \partial_* h_*$.

2

Affine Spaces

By a (real) *affine space* E is meant a non-empty set E, whose elements are called points, together with a transformation ϕ of $E \times E$ into a real vector space V, for which

(1) $\phi(P, Q) + \phi(Q, R) = \phi(P, R)$.

(2) To each $x \in V$ and $P \in E$ there exists exactly one $Q \in E$ such that $\phi(P, Q) = x$.

Intuitively, $\phi(P, Q)$ means a vector with beginning point P and end point Q. It is easy to see that $\phi(P, P) = 0$ and $\phi(P, Q) = -\phi(Q, P)$. Let E' also be an affine space and let $\phi' : E' \times E' \longrightarrow V'$ be the corresponding transformation. Then E is called a *subspace* of E' if and only if $E \subset E'$, $V \subset V'$, and $\phi(P, Q) = \phi'(P, Q)$ for all $P, Q \in E$.

Let E be a subspace of E'. Select a point O' of E' and call it the *origin*. Associate with each $P \in E$ the vector $\phi'(O', P)$, which will be termed the *coordinate vector* of P with respect to O'. By 2, a change of origin to $O'' \in E$ means the addition to each coordinate vector of the fixed vector $\phi'(O'', O')$. If P has the coordinate vector x, and Q has the coordinate vector y, then

$$\phi(P, Q) = \phi'(P, O') + \phi'(O', Q) = y - x.$$

The following construction is independent of the choice of O': Let $P_1, P_2, \cdots, P_r$ be points of E, and let $x_1, x_2, \cdots, x_r \in V'$ be their coordinate vectors with respect to $O' \in E'$. Let $\xi_1, \xi_2, \cdots, \xi_r$ be real numbers with $\sum \xi_i = 1$. The vector $x = \sum \xi_i x_i$ is coordinate vector of a unique point $S \in E'$. If $a \in V'$, a replacement of each x_i by $x_i + a$ means the replacement of x by $x + a$. Hence, S is independent of the choice of O'. If, in particular, O' is chosen in E, then all the x_i are in V, and thus, $x \in V$. This means that the constructed point S lies in E.

As an example, consider the two distinct points P and Q with coordinate vectors x and y. Let $\xi_1 = t$ and $\xi_2 = 1 - t$. The construction yields $tx + (1 - t)y = y + t(x - y)$ as coordinate vector of S. If t runs through all the real numbers, then the set of corresponding points S is termed the *line* through P and Q. It lies entirely in the subspace E, and is itself a subspace.

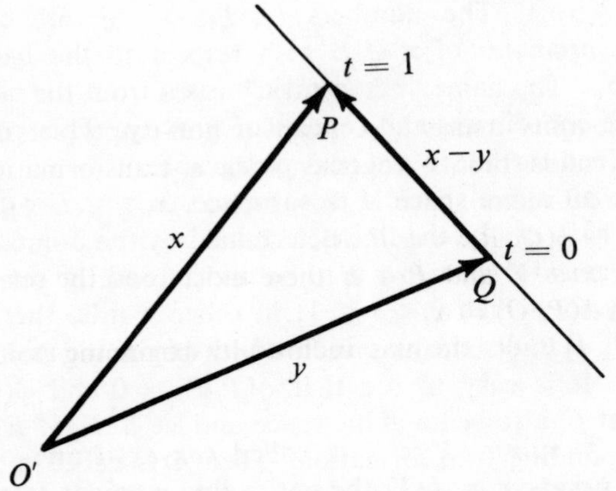

Figure 2

The *dimension* of E is defined by $\dim E = \dim V$. The $q + 1$ points with coordinate vectors $x_0, x_1, \cdots, x_q$ are called *independent* if and only if the q vectors $x_1 - x_0, x_2 - x_0, \cdots, x_q - x_0$ are linearly independent. This concept is independent of the choice of O' and the choice of x_0, as can be seen from

$$x_i - x_0 = (x_i - x_j) - (x_0 - x_j).$$

From now on, for a fixed choice of O', the points of E and the corresponding coordinate vectors will be designated by the same letters. Let E have the dimension q and let the $q + 1$ points $x_0, x_1, \cdots, x_q$ be independent points of E. Then $x_1 - x_0, \cdots, x_q - x_0$ form a basis of V. Consequently, for each $x \in E$, the vector $x - x_0$ can be written in the form

$$x - x_0 = \sum_{i=1}^{q} \xi_i(x_i - x_0),$$

with uniquely determined real ξ_i. Let $\xi_0 = 1 - \sum_{i=0}^{q} \xi_i$. Then the representation $x = \sum_{i=0}^{q} \xi_i x_i$ has uniquely determined real coefficients ξ_i with $\sum_{i=0}^{q} \xi_i = 1$; consequently, the points $x_0, x_1, \cdots, x_q$ are said to *span* E. The representation has already been proved to be independent of the choice of origin. It is thus possible to take $\xi_0, \xi_1, \cdots, \xi_q$ as coordinates of x. This yields an embedding of E in the $(q + 1)$-dimensional affine space with coordinates $\xi_0, \xi_1, \cdots, \xi_q$ such that E becomes the hyperplane which is described by the equation $\sum \xi_i = 1$. The numbers $\xi_0, \xi_1, \cdots, \xi_q$ are called the *barycentric coordinates* of $x \in E$ with respect to the base points $x_0, x_1, \cdots, x_q$. The name "barycentric" arises from the connection between these coordinates and centers of gravity. This connection will be described further in the next paragraph.

Let E' be an affine space, E a subspace, $x, y \in E$ with $x \neq y$, and let $tx + (1 - t)y$ be the line determined by the points x and y. By the *line segment from x to y* is to be understood the set of points $tx + (1 - t)y$ for which $0 \leq t \leq 1$; in other words, the image of the interval $[0, 1]$ under the map induced by t from the real numbers to the line.

Definition: A subset $X \subset E$ is called *convex* if and only if to each pair of points $x, y \in X$, the entire line segment from x to y is in X.

Example: Let $a_1, a_2, \cdots, a_r$ be points of E. Let X consist of all the points of the form $x = \sum_{i=1}^{r} \xi_i a_i$ for which $\sum \xi_i = 1$ and $\xi_i \geq 0$. Let $y \in X, y = \sum \eta_i a_i$ and let $0 \leq t \leq 1$. For a point

$$tx + (1 - t)y = \sum (t\,\xi_i + (1 - t)\,\eta_i)a_i$$

of the line segment from x to y, it then follows that

$$t\,\xi_i + (1 - t)\eta_i \geq 0$$

and

$$\sum (t\xi_i + (1 - t)\eta_i) = t + 1 - t = 1.$$

Hence, $tx + (1 - t)y \in X$ for $0 \leq t \leq 1$, and X is convex.

Let $\mu_1, \mu_2, \cdots, \mu_r$ be positive real numbers for which $\mu = \mu_1 + \mu_2 + \cdots + \mu_r > 0$. By "the center of gravity of the points a_i with mass distribution μ_i" is meant the point $S = \sum_{i=1}^{r} (\mu_i/\mu)a_i$. Here $\mu_i/\mu \geq 0$ and $\sum (\mu_i/\mu) = 1$. Clearly, the convex set of the example is just the set of the centers of gravity obtained from all possible mass distributions.

2.1 Lemma: *Let Y be a convex set and $a_1, a_2, \cdots, a_r \in Y$. Then each center of gravity of $a_1, a_2, \cdots, a_r$ is a point of Y.*

The proof is carried out by induction on r. The case $r = 1$ is trivial. Suppose now that the lemma has been proved for $r - 1$. Let $\mu_1, \mu_2, \cdots, \mu_r$ be a mass distribution with at least one non-zero μ_i, say, $\mu_1 > 0$. Let s' be the center of gravity of $a_1, a_2, \cdots, a_{r-1}$ with the mass distribution $\mu_1, \mu_2, \cdots, \mu_{r-1}$, for which

$$\mu' = \mu_1 + \mu_2 + \cdots + \mu_{r-1}.$$

Then $s = (\mu'/\mu)s' + (\mu_r/\mu)a_r$, or $s = ts' + (1 - t)a_r$, where $t = \mu'/\mu$ and $0 \leq t \leq 1$. By inductive hypothesis, $s' \in Y$; and hence, by the definition of convexity, $s \in Y$.

By the *convex hull of a set* $X \subset E$ is meant the smallest convex subset of E that contains X. Its existence is trivial since the intersection of convex sets is convex and E is convex.

2.2. THEOREM: *The convex hull of r points $a_1, a_2, \cdots, a_r$ of E consists precisely of all the points of the form $x = \sum \xi_i a_i$, where $\sum \xi_i = 1$ and $\xi_i \geq 0$.*

Proof: This set is convex; and, according to Lemma 3, is contained in every convex set that contains the points $a_1, a_2, \cdots, a_r$.

The convex hull of $r + 1$ independent points is called a *geometrical simplex* of dimension r. The convex hull of $r + 1$ dependent points is called a *singular geometrical simplex*. The $r + 1$ points are called *vertices* of the simplex.

Let E, E' be affine spaces and let V, V' be the corresponding vector spaces. The vectors $\phi(P, Q)$ and $\phi(P', Q')$ will be written (P, Q) and (P', Q') from now on.

Definition: A function $f{:}E \longrightarrow E'$ is called *affine* (or an affine map) if and only if it has the two following properties:

(1) The equation $(P, Q) = (P_1, Q_1)$ implies $(f(P), f(Q)) = (f(P_1), f(Q_1))$ and

(2) The induced mapping $\tilde{f}{:}V \longrightarrow V'$ defined by $\tilde{f}(P, Q) = (f(P), f(Q))$ is a vector space homomorphism.

By (1), the function $\tilde{f}$ is well defined.

Actually, it is possible to formulate the definition of an affine map somewhat more weakly. For instance, it is sufficient, in place of (2), to require only that $\tilde{f}(\xi x) = \xi \tilde{f}(x)$ for $x \in V$ and real ξ. For $\tilde{f}(x + y) = \tilde{f}(x) + \tilde{f}(y)$ is obtained immediately from (1) as follows: There exists $R \in E$ with $x = (P, Q)$, $y = (Q, R)$, yielding $x + y = (P, R)$. Hence,

$$\begin{aligned}\tilde{f}(x + y) = \tilde{f}(P, R) &= (f(P), f(R)) \\ &= (f(P), f(Q)) + (f(Q), f(R)) = \tilde{f}(x) + \tilde{f}(y).\end{aligned}$$

Obviously, if $f{:}E \longrightarrow E'$ and $g{:}E' \longrightarrow E''$ are affine maps, then $gf{:}E \longrightarrow E''$ is an affine map.

Let $P_0, P_1, \cdots, P_r, P, O$ be points of E and let $OP = \sum_{i=0}^{r} \xi_i(OP_i) \in V$, where $\sum \xi_i = 1$. Since $\tilde{f}$ is a homomorphism, it follows that

$$(f(O), f(P)) = \tilde{f}(OP) = \sum_{i=0}^{r} \xi_i \tilde{f}(OP_i) = \sum_{i=0}^{r} \xi_i (f(O), f(P)).$$

For an arbitrary O' from E',

$$\begin{aligned}(O', f(P)) &= (O', f(O)) + (f(O), f(P)) \\ &= \sum_{i=0}^{r} \xi_i (O', f(O)) + \sum_{i=0}^{r} \xi_i (f(O), f(P_i)),\end{aligned}$$

and thus,

$$(O', f(P)) = \sum_{i=0}^{r} \xi_i (O', f(P_i)).$$

Let $x_0, x_1, \cdots, x_r, x$ be the coordinate vectors of $P_0, P_1, \cdots, P_r, P$ with respect to $O \in E$ and let $f(x_0), f(x_1), \cdots, f(x_r), f(x)$ be the coordinate vectors of $f(P_0), f(P_1), \cdots, f(P_r)$ and $f(P)$ with respect to $O' \in E'$. This proves

2.3. *Let* $x = \sum_{i=0}^{r} \xi_i x_i$, *where* $\sum \xi_i = 1$. *Then this relation holds with the same* ξ_i *for each choice of* O *and implies the relation* $f(x) = \sum \xi_i f(x_i)$ *for any choice of* O$'$.

From this follows immediately

2.4. THEOREM: *The affine image of an r-dimensional geometrical simplex is a (possibly singular) geometrical simplex.*

In the case $r = 2$, statement 2.3 becomes

2.5. $f(tx + (1 - t)y) = tf(x) + (1 - t)f(y)$ for t real.

Thus, 2.5 is a consequence of (1) and (2) of the definition of *affine*. To prove the converse, let $f{:}E \longrightarrow E'$ be a function which satisfies 2.5 for some choice of coordinates in E and E'. Here, x and y denote the coordinate vectors of points P and Q of E, and $f(x)$ and $f(y)$ are the coordinate vectors of their images. Then $(P, Q) = y - x$, and similarly $(P_1, Q_1) = y_1 - x_1$. Proof of (1): From $(P, Q) = (P_1, Q_1)$; that is, $y - x = y_1 - x_1$, it follows that $\frac{1}{2}x + \frac{1}{2}y_1 = \frac{1}{2}x_1 + \frac{1}{2}y$. From 2.5, it follows that

$$\tfrac{1}{2}f(x) + \tfrac{1}{2}f(y_1) = f(\tfrac{1}{2}x + \tfrac{1}{2}y_1) = \tfrac{1}{2}f(x_1) + \tfrac{1}{2}f(y),$$

and hence, that $f(y) - f(x) = f(y_1) - f(x_1)$. Thus, (1) is proved. Furthermore, the function $\tilde{f}{:}V \longrightarrow V'$, given by

$$\tilde{f}(P, Q) = (f(P), f(Q)) = f(y) - f(x)$$

is well defined. For (2), it suffices to prove that $\tilde{f}(\xi x) = \xi \tilde{f}(x)$. From 2.5, for $y = 0 \in V$, it follows that

$$\begin{aligned}\tilde{f}(\xi x) &= f(\xi x) - f(0)\\ &= \xi f(x) + (1 - \xi)f(0) - f(0) = \xi f(x) - \xi f(0) = \xi \tilde{f}(x).\end{aligned}$$

Condition 2.5 implies that the affine image of a convex subset of an affine space is convex, for the line segment between x and y has as affine image the line segment between $f(x)$ and $f(y)$.

2.6. THEOREM: *Let E and E$'$ be affine spaces, with* $\dim E = q$.

Let σ be a q-dimensional simplex having vertices P_i with coordinate vectors a_i, $i = 0, 1, \cdots, q$. Let P'_i, $i = 0, 1, \cdots, q$, be points of E' whose corresponding coordinate vectors are b_i. Then there is precisely one affine map $f{:}E \longrightarrow E'$ for which $f(P_i) = P'_i$.

Proof: Assume that f is affine, and that $f(P_i) = P'_i$, $i = 0, 1, \cdots, q$. Each coordinate vector x has a unique representation $x = \sum \xi_i a_i$ with real coefficients ξ_i for which $\sum \xi_i = 1$. By 2.3, it follows that $f(x) = \sum \xi_i f(a_i)$. Thus, $f(x)$ is uniquely determined by the vectors $f(a_i)$. To prove the existence of f, it is therefore sufficient to let $f(x) = \sum \xi_i b_i$ and to prove 2.5. From

$$tx + (1-t)y = t\sum \xi_i a_i + (1-t)\sum \eta_i a_i = \sum (t\xi_i + (1-t)\eta_i)a_i,$$

it follows that

$$\begin{aligned} f(tx + (1-t)y) &= \sum (t\xi_i + (1-t)\eta_i)b_i \\ &= t\sum \xi_i b_i + (1-t)\sum \eta_i b_i \\ &= tf(x) + (1-t)f(y). \end{aligned}$$

The affine map f just defined sends P_i on P'_i, for each i, as required. Since the ξ_i are preserved, the image of σ under f is the possibly singular simplex whose vertices are P'_i.

3

Affine Simplices and Boundary Operator

A few preparatory remarks are needed so that a suitable formalism can be set up to handle affine mappings of geometrical simplices. For this purpose, it should be remembered that R is a fixed ring with unit element, and that all R-modules are unitary. Let I be an index set and M_i, $i \in I$, be R-modules. The set of all functions on I with values $f(i) \in M_i$, where $i \in I$, is called the *direct product* of the modules M_i; it is denoted by $M = \prod M_i$. The elements of M are written $x = (\cdots, x_i, \cdots)$. M becomes an R-module upon the introduction of the operations $x + y = (\cdots, x_i + y_i, \cdots)$ and $rx = (\cdots, rx_i, \cdots)$, where $r \in R$. The set of all $(\cdots, x_i, \cdots)$ in which $x_i = 0$ for all but a finite set of indices i is a submodule S of M called the *direct sum* of the modules M_i, and is written $S = \sum M_i$ or $S = \sum^{\oplus} M_i$. The elements x_i and $(0, \cdots, 0, x_i, 0, \cdots, 0)$ are identified. Consequently, $x = \sum x_i$ for $x \in S$. When I is finite, $S = M$. Otherwise, M has a higher cardinality than S.

Definition: By a *generating system* of an R-module A is meant a subset A_0 of A such that each $x \in A$ can be written in the form

$$x = \sum_{\alpha \in A_0} r_\alpha \alpha, \text{ with } r_\alpha \in R \text{ and all but a finite set of } r_\alpha \text{ zero.}$$

If each $x \in A$ has only one such representation, A_0 is called a *basis* of A. A module which has a basis is called *free*. That not all modules are free can be seen from the example of an arbitrary finite group taken as a Z-module.

If A is a free R-module with basis A_0 and $\alpha \in A_0$, then $R\alpha$ and R are isomorphic as modules because they are free. Here, A is isomorphic to the direct sum of a set of submodules, each isomorphic to R, which are in number the cardinality of A_0. A can be identified with this sum and written

$$A = \sum_{\alpha \in A_0} R\alpha.$$

Conversely, from each set A_0, a free module A can be constructed with A_0 as basis: To each $\alpha \in A_0$ associate a module R_α isomorphic to R, and define A as the direct sum of all R_α, $\alpha \in A_0$. Designate by $r\alpha$ the image of $r \in R$ under the isomorphism of R on R_α. Then $r_1\alpha = r_2\alpha$ if and only if $r_1 = r_2$. Furthermore, $(r_1 + r_2)\alpha = r_1\alpha + r_2\alpha$; $r_1({}_2\alpha) = (r_1r_2)\alpha$; and A consists of all finite sums $\sum_{\alpha \in A_0} r_\alpha \alpha$. A is called the R-module generated by the basis A. It is trivial to prove

3.1a. THEOREM: *Let $A = \sum_{\alpha \in A_0} R\alpha$ be a free R-module with basis A_0. Let $\{b_\alpha\}$ be a subset of an R-module B indexed by A_0. Then there is precisely one R-homomorphism $f{:}A \longrightarrow B$ for which $f(\alpha) = b_\alpha$ for $\alpha \in A_0$.*

Let A, B, and C be arbitrary R-modules. A function $f{:}A \times B \longrightarrow C$ is called *bilinear* if and only if

$$f(a + a', b) = f(a, b) + f(a', b)$$
$$f(a, b + b') = f(a, b) + f(a, b')$$

and

$$f(ra, b) = rf(a, b) = f(a, rb)$$

for $a \in A$, $b \in B$, $r \in R$. If f is interpreted as a composition and $f(a, b)$ is written as the product ab, then bilinearity implies distributivity from both sides and the homogeneity condition

$$(ra)b = r(ab) = a(rb).$$

It is trivial to prove

3.1b. THEOREM: *Let A, B be free R-modules with bases A_0 and B_0, and let C be an arbitrary R-module. Let $\{c_{\alpha\beta}\}$ be a subset of C indexed by $A_0 \times B_0$. Then there is a unique bilinear function $f: A \times B \longrightarrow C$ such that $f(\alpha, \beta) = c_{\alpha, \beta}$ for $\alpha \in A_0$, $\beta \in B_0$.*

Let A, B, and C be free R-modules for whose elements there are defined distributive and homogeneous products ab, bc, $(ab)c$, $a(bc)$ [with values in certain modules, not of any interest at the moment]. Then it is easy to see that the associative law $a(bc) = (ab)c$ holds precisely when it is satisfied by the basis elements.

In this volume, the composition will always be of the following sort: Let X, Y, and Z be sets. Let the free module A have as basis certain functions sending Y in Z, and let the free module B have as basis certain functions sending X in Y. The successive application of these functions defines a composition ab whose values lie in a module generated by certain functions sending X in Z. The associative law holds for functions and, therefore, for this composition.

Special symbols are used for particular choices of the spaces and functions: When X and Y are topological spaces, then the free module generated by all continuous maps of X in Y is denoted by $C(X, Y)$. If X or Y is empty, $C(X, Y)$ is defined to be 0.

Let X and Y be convex subsets of affine spaces. A map of X in Y is called *affine* if and only if it is obtained from an affine map of the affine spaces by restricting the domain to X and the image to Y. A map $f: X \longrightarrow Y$ is thus affine precisely when it satisfies 2.5 of Chapter 2. The set of affine maps of X in Y will be denoted by $A_0(X, Y)$. The free R-module generated by all affine maps of X in Y will be denoted by $A(X, Y)$; it has $A_0(X, Y)$ as a basis. If X or Y is empty, $A(X, Y) = 0$ by definition, as before. At the end of this chapter the affine spaces will be topologized. Thereby, $A(X, Y)$ will become a submodule of $C(X, Y)$.

Consider an affine countably infinite dimensional space whose corresponding vector space is generated by the vectors $d_0, d_1, \cdots$. The points $d_0, d_1, \cdots, d_q$ (more precisely, the points having coordinate vectors $d_1, d_2, \cdots, d_q$ with respect to the arbitrary fixed point O) span a q-dimensional subspace whose points have the coordinate vectors $x = \sum_{i=0}^{q} \xi_i d_i$, where $\sum \xi_i = 1$. The convex hull of the points $d_0, d_1, \cdots, d_q$ is a q-dimensional geometrical simplex; it will be denoted by Δ_q.

It has been proved that there is a one-to-one correspondence between the points of Δ_q and their barycentric coordinates $(\xi_0, \xi_1, \cdots, \xi_q)$, where $\xi_i \geq 0$ and $\sum \xi_i = 1$. This correspondence was independent of the choice of O. Therefore, Δ_q can also be defined as follows: Consider an arbitrary set of $q + 1$ objects $\{d_0, d_1, \cdots, d_q\}$ and all the maps of this set in the real numbers such that, for the images ξ_i of d_i, both $\xi_i \geq 0$ and $\sum \xi_i = 1$. The individual maps are called points of Δ_q, and Δ_q is the collection of all its points. Let E be a convex subset of an affine space and let $a_0, a_1, \cdots, a_q$ be points of E. The affine map $f{:}\Delta_q \longrightarrow E$ for which $f(d_i) = a_i$, for $i = 0, 1, \cdots, q$, is unique by Theorem 2.6 and will be denoted by $(a_0, a_1, \cdots, a_q)$. The map $(a_0, a_1, \cdots, a_q)$ will be called the *affine simplex of E* with the (ordered) vertices $a_0, a_1, \cdots, a_q$. By Theorem 2.6, the image of Δ_q under $(a_0, a_1, \cdots, a_q)$ is the (possibly singular) geometrical simplex with the vertices $a_0, a_1, \cdots, a_q$. This proves

3.2. THEOREM: *The affine simplices of E form a basis for the free R-module $A(\Delta_q, E)$.*

So far, $A(\Delta_q, E)$ has been defined as a free module for $q = 0, 1, 2, \cdots$. Purely formally, the symbol $A(\Delta_q, E)$ is defined, for $q = -1$, as the free module with a basis consisting of precisely one element, which is to be denoted by $(.)$. Thus, $A(\Delta_{-1}, E) = R(.)$. The element $(.)$ is called the *empty simplex*.

Now, two homomorphisms will be introduced that are used in all auxiliary constructions but not in the principal results. Unnecessary parentheses will be omitted when they are clear from the context.

Let E and F be convex subsets of affine spaces and $\rho \in A_0(E, F)$; that is, ρ is an affine mapping of E in F. The image of $x \in E$ under

ρ will be denoted by $x^\rho = \rho(x)$. For $q \geq 0$ and $(a_0, a_1, \cdots, a_q) \in A_0(\Delta_q, E)$, this means that

$$\rho(a_0, a_1, \cdots, a_q) = (a_0^\rho, a_1^\rho, \cdots, a_q^\rho) \in A_0(\Delta_q, F).$$

For $q = -1$, $\rho\,(.)$ is defined to be $(.)$. Thus, there is a composition between the basis elements of $A(E, F)$ and those of $A(\Delta_q, E)$, with values in $A(\Delta_q, F)$. This composition is unique by Theorem 3.1b. The map ρ can also be interpreted as a module homomorphism $\rho: A(\Delta_q, E) \longrightarrow A(\Delta_q, F)$.

By Theorem 3.1a, each $x \in X$ can be interpreted as a homomorphism of $A(\Delta_q, E)$ in $A(\Delta_{q+1}, E)$ by means of the definition

$$x(a_0, a_1, \cdots, a_q) = (x, a_0, a_1, \cdots, a_q) \qquad \text{for } q \geq 0$$
$$x(.) = (x) \qquad \text{for } q = -1.$$

Simple properties of these two homomorphisms will now be found:

(1) For $\rho \in A_0(E, F)$, $\rho x = x^\rho \rho$. Here,

$$A(\Delta_q, E) \xrightarrow{x} A(\Delta_{q+1}, E) \xrightarrow{\rho} A(\Delta_{q+1}, F)$$
$$A(\Delta_q, E) \xrightarrow{\rho} A(\Delta_{q+1}, F) \xrightarrow{x} A(\Delta_{q+1}, F).$$

It is sufficient to show that the image of a basis element is the same in both cases. Clearly,

$$\rho x(a_0, a_1, \cdots, a_q) = (x^\rho, a_0^\rho, \cdots, a_q^\rho) = x^\rho \rho(a_0, a_1, \cdots, a_q).$$

(2) Let ϕ be a homomorphism of $A(\Delta_q, E)$ into an arbitrary module B. Then ϕx is a homomorphism of $A(\Delta_{q-1}, E)$ into B by means of

$$A(\Delta_{q-1}, E) \xrightarrow{x} A(\Delta_q, E) \xrightarrow{\phi} B.$$

This leads to

3.3. Lemma: *The relation $\phi x = 0$, for all x, implies that $\phi = 0$.*

Proof: By Theorems 3.1a and 3.2, it suffices to prove that $\phi(a_0, a_1, \cdots, a_q) = 0$ for all affine simplices of E. This follows from

$$\phi(a_0, a_1, \cdots, a_q) = \phi a_0(a_1, \cdots, a_q) = 0,$$

which, in turn, holds since it is true for all $a_0 \in E$.

3.4. Lemma: *Let E and F be convex subsets of affine spaces. Let $g_x:A(\Delta_q, E) \longrightarrow A(\Delta_p, F)$, where $p \geq 0$, and $h:A(\Delta_q, E) \longrightarrow A(\Delta_{p+1}, F)$, where $p \geq -1$, be homomorphisms, where the homomorphism g_x can vary with x. Furthermore, let λ be an affine map of E into F. Then there is exactly one homomorphism $f:A(\Delta_{q+1}, E) \longrightarrow A(\Delta_{q+1}, F)$ such that $fx = h + x^\lambda g_x$ for all $x \in E$.*

Proof: By Theorem 3.1a, it suffices to define f on a basis of $A(\Delta_{q+1}, E)$. Let

$$f(a_0, a_1, \cdots, a_{q+1}) = h(a_1, a_2, \cdots, a_{q+1}) + a_0^\lambda g_{a_0}(a_1, a_2, \cdots, a_{q+1}).$$

Then, for each basis element $(a_0, a_1, \cdots, a_q)$ of $A(\Delta_q, E)$ and each $x \in E$, the definition implies that

$$\begin{aligned} f_x(a_0, a_1, \cdots, a_q) &= f(x, a_0, \cdots, a_q) \\ &= h(a_0, \cdots, a_q) + x^\lambda g_x(a_0, \cdots, a_q). \end{aligned}$$

Since this holds for each affine simplex, it follows that $fx = h + x^\lambda g_x$. If there were an f_1 with $f_1 x = h + x^\lambda g_x$, then $(f - f_1) = 0$, for all $x \in E$. By Lemma 3.3, this would imply that $f_1 = f$, so that f is uniquely determined.

The Boundary Operator ∂_q: A homomorphism $\partial_q:A(\Delta_q, E) \longrightarrow A(\Delta_{q-1}, E)$ will be defined recursively, for $q \geq 0$, as follows:

$$\partial_0 x = \text{Identity on } A(\Delta_{-1}, E), \text{ and}$$

3.5.

$$\partial_{q+1}x + x\partial_q = \text{Identity on } A(\Delta_q, E) \text{ for } q \geq 0.$$

This definition is consistent since

$$A(\Delta_q, E) \xrightarrow{\partial_q} A(\Delta_{q-1}, E) \xrightarrow{x} A(\Delta_q, E)$$

and

$$A(\Delta_q, E) \xrightarrow{x} A(\Delta_{q+1}, E) \xrightarrow{\partial_{q+1}} A(\Delta_q, E).$$

Suppose that ∂_q has already been found and is unique. From Lemma 3.4 it follows, for $q = p + 1$, $\lambda =$ Identity, $h =$ Identity, and $g_x = -\partial_q$, that $\partial_{q+1}: A(\Delta_{q+1}, E) \longrightarrow A(\Delta_q, E)$ is a uniquely determined homomorphism.

It is worth knowing ∂_q explicitly, particularly for the basis elements of $A(\Delta_q, E)$:

For $q = 0$, $\partial_0(a_0) = \partial_0 a_0(.) = \text{Id}(.) = (.)$.

For $q = 1$, $\partial_1(a_0, a_1) = \partial_1 a_0(a_1) = (\text{Id} - a_0\partial_0)(a_1)$
$= (a_1) - a_0(.) = (a_1) - (a_0)$.

It should be noticed that the portion of the definition of ∂_q which reads: $\partial_0 x =$ Identity on $A(\Delta_{-1}, E)$ could be replaced by $\partial_0 x = 0$ on $A(\Delta_{-1}, E)$ to yield a new boundary operator. This would change the explicit calculation on basis elements for the case $q = 0$ to $\partial_0(a_0) = 0$, but would not affect calculations for $q > 0$. The relation $\partial_0(a_0) = 0$ can be interpreted intuitively as saying that the boundary of a point is zero. If the change in definition is maintained, it becomes necessary for the preservation formula 3.6, which follows in the next paragraph, to make the convention that $(.) = 0$. The reader can, if he cares to, follow the changes in the behavior of the case $q = 0$ that result from the change in definition. They will be ignored here until the introduction of singular homology in Chapter 4.

When $q \geq 0$, the symbol $(a_0, a_1, \cdots, \hat{a}_i, \cdots, a_q)$ will denote the $(q - 1)$-dimensional affine simplex $(a_0, a_1, \cdots, a_{i-1}, a_{i+1}, \cdots, a_q)$, obtained from the affine simplex $(a_0, a_1, \cdots, a_q)$ by deleting the vertex point a_i. Then it follows, for $q \geq 0$, that

3.6. $$\partial_q(a_0, \cdots, a_q) = \sum_{i=0}^{q} (-1)^i (a_0, \cdots, \hat{a}_i, \cdots, a_q).$$

Proof: This has already been checked for $q = 0, 1$. The proof now proceeds by complete induction: Suppose that 3.6 has already been proved for $q - 1$. Then,

$$\begin{aligned}\partial_q(a_0, \cdots, a_q) &= \partial a_0(a_1, \cdots, a_q) = (\text{Id} - a_0\partial_{q-1})(a_1, \cdots, a_q)\\ &= (a_1, \cdots, a_q) - a_0 \sum_{i=0}^{q} (-1)^{i-1}(a_1, \cdots, \hat{a}_i, \cdots, a_q)\\ &= \sum_{i=0}^{q} (-1)^i (a_0, \cdots, \hat{a}_i, \cdots, a_q).\end{aligned}$$

Formula 3.6 implies that, apart from signs, the boundary of an affine simplex is the sum of its boundary simplices.

3.7. *For $x \in E$ and $q \geq 0$, $\partial_q x \partial_q = \partial_q$, and $\partial_q \partial_{q+1} = 0$.*

Proof: By definition, $\partial_0(a_0) = (.)$, and

$$\partial_0 x \partial_0(a_0) = \partial_0 x(.) = \partial_0(x) = (.)$$

Suppose that the formula $\partial_q x \partial_q = \partial_q$ has been proved for some $q \geq 0$. Multiplication of $\partial_{q+1}x + x\partial_q = \text{Id}$ on the left by ∂_q yields $\partial_q\partial_{q+1}x + \partial_q x \partial_q = \partial_q$, and thus $\partial_q\partial_{q+1}x = 0$ for all x. Lemma 3.3 implies that $\partial_q\partial_{q+1} = 0$. Now multiply $\partial_{q+1}x + x\partial_q = \text{Id}$ on the right by ∂_{q+1} to obtain $\partial_{q+1}x\partial_{q+1} + x\partial_q\partial_{q+1} = \partial_{q+1}$; that is, $\partial_{q+1}x\partial_{q+1} = \partial_{q+1}$.

The composition between $A(E, F)$ and $A(\Delta_q, E)$ with values in $A(\Delta_q, F)$ was taken (Theorem 3.1b) to be the one defined through linear combinations of the "products" obtained by writing in succession a basis element of $A(E, F)$ and a basis element of $A(\Delta_q, E)$. These basis elements are mappings $\Delta_q \longrightarrow E$ and $E \longrightarrow F$, respectively. In the special case where $E = \Delta_q$ and $q \geq 1$, let $(d_0, \cdots, \hat{d}_i, \cdots, d_q)$ denote a particular basis element of $A(\Delta_{q-1}, \Delta_q)$. Then, for $a_0, a_1, \cdots, a_q \in E$ and $q \geq 1$,

3.8. $(a_0, a_1, \cdots, a_q)(d_0, \cdots, \hat{d}_i, \cdots, d_q) = (a_0, \cdots, \hat{a}_i, \cdots, a_q).$

Proof: Both sides of the equation are affine maps $\Delta_{q-1} \longrightarrow E$. By Theorem 2.6, these are equal if they yield the same images of a set of q independent points, say $d_0, d_1, \cdots, d_q$. The map $(d_0, d_1, \cdots, \hat{d}_i, \cdots, d_q)$ sends the point d_ν on d_ν for $\nu < i$ and on $d_{\nu+1}$ for $\nu \geq i$, $\nu = 0, 1, \cdots, q-1$. The point d_ν is thus sent by

$$(a_0, a_1, \cdots, a_q)(d_0, d_1, \cdots, \hat{d}_i, \cdots, d_q)$$

on a_ν when $\nu < i$ and on $a_{\nu+1}$ when $\nu \geq i$. The map $(a_0, \cdots, \hat{a}_i, \cdots, a_q)$ has the same effect.

Notation: In the module $A(\Delta_q, \Delta_q)$, the map $(d_0, d_1, \cdots, d_q)$ is the identity. It will be denoted by $\Delta_q^\dagger$.
For $q \geq 1$ and $c \in A(\Delta_q, E)$,

3.9. $$c(\partial_q \Delta_q^\dagger) = \partial_q c.$$

Notice that on the left the module element $c \in A(\Delta_q, E)$ and $\partial_q \Delta_q^\dagger \in A(\Delta_{q-1}, \Delta_q)$ are composed. Property 3.9 is important; for it says that the identity of $A(\Delta_q, \Delta_q)$ is the only element for which ∂_q needs to be computed.

Proof of 3.9: Multiply 3.8 by $(-1)^i$ and sum over i to obtain

$$(a_0, a_1, \cdots, a_q) \sum_{i=0}^{q} (-1)^i (d_0, \cdots, \hat{d}_i, \cdots, d_q)$$
$$= \sum_{i=0}^{q} (-1)^i (a_0, \cdots, \hat{a}_i, \cdots, a_q)$$

or

$$(a_0, a_1, \cdots, a_q)\partial_q \Delta_q^\dagger = \partial_q(a_0, a_1, \cdots, a_q).$$

Now 3.9 follows for each $c \in A(\Delta_q, E)$ by the taking of linear combinations.

For $q \geq 2$, a twofold application of 3.9 yields

$$0 = \partial_{q-1}\partial_q(c) = \partial_{q-1}(c\partial_q(\Delta_q^\dagger)) = c\partial_q(\Delta_q^\dagger)\partial_{q-1}(\Delta_{q-1}^\dagger).$$

The last expression arises in the composition of three modules; and parentheses are omitted since mappings, and thus module compositions, are associative. Now specialize E to Δ_q and c to

$$c = \Delta_q^\dagger \in A(\Delta_q, \Delta_q).$$

Then,

$$\partial_{q-1}\Delta_{q-1}^\dagger \in A(\Delta_{q-2}, \Delta_{q-1})$$

and

$$\partial_q \Delta_q^\dagger \in A(\Delta_{q-1}, \Delta_q).$$

Since the factor $\Delta_q^\dagger = \text{Id}$ is superfluous, it follows that

3.10. $$\partial_q(\Delta_q^\dagger)\partial_{q-1}(\Delta_{q-1}^\dagger) = 0, \text{ for } q \geq 2.$$

The Topologization of Affine Spaces: An affine space E of finite dimension n is metrized as follows: Let $e_1, e_2, \cdots, e_n$ be a basis for the corresponding vector space, and define $|a|$, for $a = \sum a_i e_i \in V$, by $|a| = \sqrt{\sum a_i^2}$. Then V becomes a normed vector space. If P and Q are points of E with coordinate vectors x and y, then $d(P, Q)$ is defined by $d(P, Q) = |y - x|$. It is easy to check that this makes E a metric space. The metric depends upon the choice of the basis

$e_1, e_2, \cdots, e_n$, but the topology does not because each spherical ball defined in terms of one basis contains one defined by each arbitrarily chosen basis. (The reader should consider this statement as an exercise to be checked.) Subspaces, such as Δ_q, always have the induced topology. The metric is introduced in infinite dimensional affine spaces by analogous devices, but here the topology is a function of the metric.

If Δ_q is embedded in the $(q + 1)$-dimensional affine space that was defined by the barycentric coordinates $\xi_0, \xi_1, \cdots, \xi_q$, then the complement of Δ_q consists of the points for which one of the conditions $\xi_i \geq 0$, $\sum \xi_i = 1$ fails. The complement is therefore open; hence, Δ_q is closed. Since E is homeomorphic to R^{q+1}, and since Δ_q is bounded, it follows that Δ_q, and thus its continuous images, are compact. Each affine map is linear and therefore continuous; hence, each geometrical simplex is compact.

4

The Singular Homology Theory

Let X be a topological space. Since Δ_q is also a topological space, it is possible to define the free R-module $C(\Delta_q, X)$, $q \geq 0$, generated by the set of all continuous maps $\sigma \colon \Delta_q \longrightarrow X$. Each such map σ is called a *singular q-simplex* of X. To each integer q, a free R-module $S_q(X)$, whose elements are called *chains* (or q-chains), is defined by

4.1. $$S_q(X) = \begin{cases} C(\Delta_q, X) & \text{for } q \geq 0 \\ R(.) & \text{for } q = -1 \\ 0 & \text{for } q < -1 \end{cases}, \text{ if } X \neq \varnothing$$

$$S_q(X) = 0 \text{ for all } q \in Z, \text{ if } X = \varnothing.$$

Here, $\varnothing$ denotes the empty set. For $X \neq \varnothing$ and $q \geq 0$, $S_q(X) \neq 0$, since Δ_q can be mapped continuously on a single point of X. When $X \neq \varnothing$ and $q \geq 0$, each $c \in S_q(X)$ is associated with a set $|c|$ in X called the *carrier* of c. Intuitively, the carrier is the image set; it is defined by

$$|c| = \bigcup_{r_\sigma \neq 0} \sigma(\Delta_q) \qquad \text{for } c = \sum r_\sigma \sigma \neq 0$$
$$|c| = \varnothing \qquad \text{for } c = 0.$$

It will be proved that the R-modules $S_q(X)$ form a chain complex when the boundary operator ∂_q is defined for $c \in S_q(X)$ by

4.2. $$\partial_q c = \left\{\begin{array}{ll} c\partial_q(\Delta_q^\dagger) & \text{for } q \geq 1 \\ (\sum r_\sigma)(.) & \text{for } q = 0,\ c = \sum r_\sigma \sigma \\ 0 & \text{for } q \leq -1, \end{array}\right\}, \text{ if } X \neq \varnothing$$

and by

$$\partial_q c = 0 \quad \text{for all } q, \quad \text{if } X = \varnothing.$$

In this definition, the expression $c\partial_q(\Delta_q^\dagger)$ is defined by means of composition of the modules $C(\Delta_q, X)$ and $A(\Delta_{q-1}, \Delta_q)$, and has values in $C(\Delta_{q-1}, X)$. Thus, $c\partial_q(\Delta_q^\dagger) \in C(\Delta_{q-1}, X)$. Since the composition is bilinear (Theorem 3.1b), ∂_q is a homomorphism of $S_q(X)$ into $S_{q-1}(X)$. In the cases $q \leq 1$ or $X = \varnothing$, ∂_q is trivially a homomorphism. The element $\partial_q c$ is called the *boundary of the chain c*. From the definitions, it follows immediately that

$$\cdots \longrightarrow S_{q+1}(X) \xrightarrow{\partial_{q+1}} S_q(X) \xrightarrow{\partial_q} S_{q-1}(X) \longrightarrow \cdots.$$

For $q = 0$, $S_0(X)$ is the R-module whose basis elements are the continuous maps of d_0 in X. These maps can be identified with the points of X. Thus, $S_0(X)$ can also be conceived of as the free R-module generated by the points of X. The boundary of a 0-simplex; i.e., of a point in X, is then (.) by definition.

Each singular 1-simplex is a continuous map of the interval Δ_1 into X; and thus a continuous curve in X together with its parametrization. Therefore, $S_1(X)$ is the free R-module whose basis is the set of continuous curves with given parametrization. Since $\partial_1(\Delta_1^\dagger) = (d_1) - (d_0)$, a composition with $\sigma : \Delta_1 \longrightarrow X$ yields $\sigma((d_1) - (d_0)) = \sigma(d_1) - \sigma(d_0)$; that is, $\partial_1 \sigma = \sigma(d_1) - \sigma(d_0)$. This can be stated as: *The boundary of a curve σ is the difference obtained by subtracting the endpoint from the beginning point.* This difference is meaningful in the free module generated by the points of X.

By 3.6 it follows that, when c is a singular q-simplex and $q \geq 0$,

$$\partial_q \sigma = \sum_{i=0}^{q} (-1)^i \sigma(d_0, \cdots, \hat{d}_i, \cdots, d_q).$$

This formula shows that when $q \geq 1$, then $|\partial_q \sigma| \subset |\sigma|$, and more generally, that $|\partial_q c| \subset |c|$. Intuitively, then, $c \in S_q(X)$ represents, for $q \geq 0$ and $R = Z$, a surface that is an assembly of a finite number of continuous images of Δ_q, each counted with some multiplicity (possibly negative). For short, this is called *a triangulated surface in X with weights for the individual pieces.* The formula for $\partial_q c$ then states that, for $c = \sum r_\sigma \sigma$, $\partial_q c = \sum r_\sigma \partial_q \sigma$, and that the boundary of the triangulated surface is the sum of the boundaries of the pieces each counted with the weight of the piece from which it came. Here, the boundary $\partial_q \sigma$ of σ is to be interpreted so that $\sigma(d_0, \cdots, \hat{d}_i, \cdots, d_q)$ is a part of the boundary of σ counted with the weight $(-1)^i$.

4.3. THEOREM: *The sequence*

$$\cdots \longrightarrow S_{q+1}(X) \xrightarrow{\partial_{q+1}} S_q(X) \xrightarrow{\partial_q} S_{q-1}(X) \longrightarrow \cdots$$

is a chain complex.

Proof: For $X = \varnothing$, the statement is trivial. Let $X = \varnothing$. It is sufficient to prove that $\partial_{q-1}\partial_q = 0$. From 3.10 and from the associativity of the composition, it follows, for $q \geq 2$, that

$$\partial_{q-1}\partial_q c = \partial_{q-1}(c\partial_q(\Delta_q^\dagger)) = c\partial_q(\Delta_q^\dagger)\partial_{q-1}(\Delta_{q-1}^\dagger) = 0.$$

For $q \leq 0$, $\partial_{q-1} = 0$, and therefore, $\partial_{q-1}\partial_q = 0$. Finally, for $q = 1$,

$$\partial_0 \partial_1 c = \partial_0(c(\partial_1) - c(d_0)) = (1 - 1)(.) = 0.$$

For $X \neq \varnothing$, the definitions were that $S_{-1}(X) = R(.)$ and $\partial_0 c = \sum r_\sigma(.)$.

A complex containing chains of the form $r(.)$ is called *augmented*, and the homology theory of such a complex is said to be *reduced*.

There is another very useful case, the *non-augmented* one, where both $S_{-1}(X)$ and the boundaries of 0-chains are defined to be zero, while no changes are made in other dimensions. Intuitively, the non-augmented case is the one in which the boundaries of points are 0. Theorem 4.3 also holds in the non-augmented case, since $\partial_0 = 0$ implies that both $\partial_{-1}\partial_0 = 0$ and $\partial_0\partial_1 = 0$.

In both the augmented and the non-augmented cases, the chain complex of Theorem 4.3 yields a homology group as described in Chapter 1.

Let A be a subspace of X. The space A has the chain complex $\cdots \longrightarrow S_{q+1}(A) \longrightarrow S_q(A) \longrightarrow \cdots$. Since the injection $A \xrightarrow{i} X$ is continuous, each singular q-simplex of A can be taken as a singular q-simplex of X by $\Delta_q \xrightarrow{\sigma} A \xrightarrow{i} X$. This means that $c \in S_q(A)$, for a chain c of $S_q(X)$, precisely when $|c| \subset A$. Hence, $S_q(A)$ is a submodule of $S_q(X)$ for $q \geq 0$. This is also true, even trivial, when $q < 0$. More precisely, $S_q(A) = S_q(X)$ for $q < 0$, except in the augmented case when $q = -1$, $X \neq \varnothing$ and $A = \varnothing$.

Thus, to each pair A, X for which $A \subset X$ there is associated the factor module $S_q(X, A) = S_q(X)/S_q(A)$, to be read "$S_q$ of X modulo A." For $q < 0$, $S_q(X, A) = 0$, except in the augmented case when $q = -1$, $X \neq \varnothing$, and $A = \varnothing$. For this exception, $S_{-1}(X, \varnothing) = R(.)$.

For $S_q(X, A)$ the boundary operator ∂_q is defined by

$$\partial_q(c_q + S_q(A)) = \partial_q c_q + S_{q-1}(A), \text{ for } c \in S_q(X).$$

Since $\partial_q(S_q(A)) \subset S_{q-1}(A)$, Lemma 1.1 guarantees that ∂_q is well defined and a homomorphism $\partial_q : S_q(X, A) \longrightarrow S_{q-1}(S, A)$. It follows that

$$\begin{aligned}\partial_{q-1}\partial_q(c + S_q(A)) &= \partial_{q-1}(\partial_q c + S_{q-1}(A)) \\ &= \partial_{q-1}\partial_q c + S_{q-2}(A) = 0 + S_{q-2}(A).\end{aligned}$$

Since $S_{q-2}(A)$ is the zero element of $S_{q-2}(X, A)$, the homomorphism $\partial_q : S_q(X, A) \longrightarrow S_{q-1}(X, A)$ satisfies the condition $\partial_{q-1}\partial_q = 0$. Thus, the R-modules $S_q(X, A)$ form a chain complex. For $A = \varnothing$, $S_q(A) = 0$, and then $S_q(X, A) = S_q(X)$. This yields the chain complex of Theorem 4.3 as a special case.

By the definition of $S_q(X, A)$ it follows, when i is the injection and j is the canonical map on the factor module, that

$$0 \longrightarrow S_q(A) \xrightarrow{i} S_q(X) \xrightarrow{j} S_q(X, A) \longrightarrow 0$$

is exact. It follows readily from the definitions that $i\partial = \partial i$ and that $j\partial = \partial j$. Then the principal theorem of Chapter 1 can be employed to yield an exact sequence of homology groups

$$\cdots \longrightarrow H_q(A) \longrightarrow H_q(X) \longrightarrow H_q(X, A) \longrightarrow H_{q-1}(A) \longrightarrow H_{q-1}(X, A) \longrightarrow \cdots.$$

The homology groups $H_q(X, A)$ are called the *homology groups of the pair* (X, A). They are the homology groups of the chain complex $S_q(X, A)$. For $A = \varnothing$, it is customary to write $H_q(X, A) = H_q(X)$, because $S_q(X, \varnothing) = S_q(X)$.

4.4. THEOREM: *For* $q < 0$, $H_q(X, A) = 0$.

Proof: By definition, $H_q(X, A) = \text{Ker}(\partial_q)/\partial_{q+1}(S_{q+1}(X, A))$. In the case $S_q(X, A) = 0$, it follows that $\partial_q = 0$ and hence that $H_q(X, A) = 0$. There thus remains for investigation only the augmented case in which $q = -1$, $X \neq \varnothing$, and $A = \varnothing$. Since $S_{-1}(X) = R(.)$, it follows that $\longrightarrow S_0(X) \xrightarrow{\partial_0} R(.) \xrightarrow{\partial_{-1}} 0$, and then that

$$H_{-1}(X) = \text{Ker}(\partial_{-1})/\partial_0(S_0(X)) = R(.)/R(.) = 0.$$

4.5. Corollary: *Since* $\cdots \longrightarrow H_0(A) \xrightarrow{i_{0*}} H_0(x) \xrightarrow{j_{0*}} H_0(X, A) \xrightarrow{\partial_{0*}} 0 \longrightarrow 0 \longrightarrow \cdots$ *is exact,* j_{0*} *is always an epimorphism.*

An attempt will now be made to illustrate the meaning of $\partial_*: H_q(X, A) \longrightarrow H_{q-1}(A)$ geometrically. (See Figure 3.) For this purpose, q will be chosen to be 2 and X will be a subset of affine 3-space with $A \subset X$. In the illustration, A is shaded. The lower, heavily outlined vase shows a cycle $c_q \in S_q(X, A)$. For the definition of ∂_* on this cycle, it is necessary to choose a chain $b_q \in S_q(X)$ such that $j(b_q) = c_q$. In the illustration, b_q is the heavily outlined vase together with a tube. Then, $\partial b_q = i(a_{q-1})$ with $a_{q-1} \in S_{q-1}(A)$, and $\partial_*(c_q + \partial S_{q+1}(X, A)) = a_{q-1} + \partial S_q(A)$ is well defined.

Let $A_1, A_2, \cdots, A_r$ be subspaces of a topological space X, and let $B_1, B_2, \cdots, B_r$ be subspaces of a topological space Y. Any of these spaces may be the empty space, $\varnothing$. The symbol

$$f:(X, A_1, \cdots, A_r) \longrightarrow (Y, B_1, \cdots, B_r)$$

is to mean that $Y \neq \varnothing$ when $X \neq \varnothing$, and also that f induces a

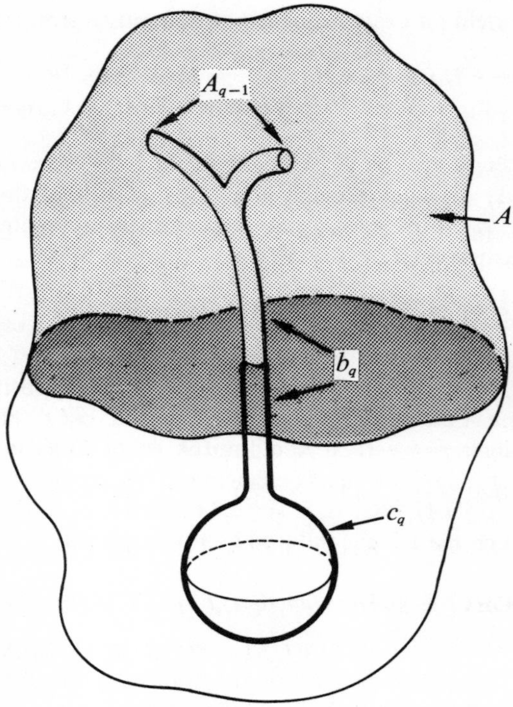

Figure 3

continuous function $g: X \longrightarrow Y$ for which $g(A_i) \subset B_i$ for $i = 1, 2, \cdots, r$. The function f is called a *continuous function of the space tuple*, and it is necessary to distinguish between f and the continuous $g: X \longrightarrow Y$. Conversely, a continuous g induces such an f if and only if $g(A_i) \subset B_i$ for $i = 1, \cdots, r$. The symbol $f: \varnothing \longrightarrow B$ means that each $x \in \varnothing$ is paired with an $f(x)$; or, in other words, that no condition is placed on f. The situation $f: \varnothing \longrightarrow B$ occurs in the restriction of a continuous function $X \longrightarrow Y$ to the empty subset of X. The pair $(X, \varnothing)$ is abbreviated X, since no misunderstandings are possible.

Examples: The identity Id: $X \longrightarrow X$ induces maps between subspaces of X. Those of the type $(A, B) \longrightarrow (X, B)$ for $B \subset A \subset X$

will be denoted by the letter i. Those of the type $(X, B) \longrightarrow (X, A)$ for $B \subset A \subset X$ will be denoted by j. In both cases, $B = \varnothing$, $A = B = \varnothing$, or $A = B = X = \varnothing$ are permissible.

Definition: Let $A \subset X$, $B \subset Y$, and let $f:(X, A) \longrightarrow (Y, B)$ be a continuous function which induces the continuous function $g: X \longrightarrow Y$. To each such f the following conditions associate an induced homomorphism $f^{\#}: S_q(X, A) \longrightarrow S_q(Y, B)$:

(1) $f_q^{\#} = 0$, if either $S_q(X, A) = 0$ or $S_q(Y, B) = 0$.

(2) $f_{-1}^{\#} = \mathrm{Id}$, if $S_{-1}(X, A) = S_{-1}(Y, B) = R(.)$. This happens only in the augmented case, and then only when $X \neq \varnothing$, $Y \neq \varnothing$, and $A = B = \varnothing$.

(3) $f_q^{\#}(c + S_q(A)) = gc + S_q(B)$ for $c \in S_q(X)$, if $X \neq \varnothing$, $Y \neq \varnothing$, and $q \geq 0$. This is the principal case.

The product gc can be formed by the composition for the modules $C(X, Y)$ and $C(\Delta_q, X)$. Hence, $gc \in S_q(Y)$. Furthermore, the function $c \longrightarrow gc$ is a homomorphism of $S(X)$ in $S(Y)$ which preserves the sum of the coefficients. This result will be needed later. If $c \in S_q(A)$; that is, if $|c| \subset A$, then $|gc| \subset B$; that is, $gc \in S_q(B)$. By Lemma 1.1, the induced homomorphism $f^{\#}$ is well defined.

4.6. THEOREM: *The function $f^{\#}: S_q(X, A) \longrightarrow S_q(Y, B)$ is a chain complex homomorphism.*

Proof: The commutativity of the following diagram is all that needs to be checked:

$$\begin{array}{ccc} S_{q-1}(X, A) & \xleftarrow{\partial_q} & S_q(X, A) \\ \downarrow f_{q-1}^{\#} & & \downarrow f_q^{\#} \\ S_{q-1}(Y, B) & \xleftarrow{\partial_q'} & S_q(Y, B); \end{array}$$

that is, $f_{q-1}^{\#}\partial_q = \partial_q' f_q^{\#}$. This is trivial in all cases for which $\partial_q = \partial_q' = 0$; namely, when $q \leq -1$, and, for non-augmented complexes, when $q = 0$. Four cases remain:

(1) Augmented complexes with $A \neq \varnothing$ or $B \neq \varnothing$ and $q = 0$: From $A \neq \varnothing$, it follows that $B \neq \varnothing$. From $B \neq \varnothing$, it follows that $S_{-1}(Y, B) = 0$, and therefore, that $\partial_0' = 0$ and $f_{-1}^{\#} = 0$. Now commutativity is trivial.

(2) Augmented complexes with $X \neq \varnothing$ and $A = B = \varnothing$ and $q = 0$: By definition, $f_{-1}^{\#} = \mathrm{Id}$; and commutativity reduces to $\partial_0 = \partial_0' f_0^{\#}$. Let $c = \sum r_\sigma \sigma \in C(\Delta_0, X)$; then it must be proved

that $\partial_0 c = \partial_0' gc$. From $gc = \sum r_\sigma g(\sigma)$, it follows that $\partial_0' gc = (\sum r_\sigma)\,(.) = \partial_0 c$.

(3) Complexes with $X = \varnothing$ or $Y = \varnothing$: Here, either all $S_q(X, A)$ or all $S_q(Y, B)$ are zero. Therefore, all $f_q^\# = 0$, from which it follows that $f_{q-1}^\# \partial_q = \partial_q' f_q^\#$ for all q.

(4) Complexes with $X \neq \varnothing$ and $Y \neq \varnothing$ when $q \geq 1$: This is the principal case. Let $c \in S_q(X)$. Then,

$$\begin{aligned} f_{q-1}^\# \partial_q(c + S_q(A)) &= f_{q-1}^\#(c\partial_q(\Delta_q^\dagger) + S_{q-1}(A)) \\ &= gc\partial_q(\Delta_q^\dagger) + S_{q-1}(B) \\ &= \partial_q'(gc + S_q(B)) \\ &= \partial_q' f_q^\#(c + S_q(A)); \end{aligned}$$

in other words, $f_{q-1}^\# \partial_q = \partial_q' f_q^\#$. The proof of the commutativity thus follows readily from the associativity of the compositions.

As was proved earlier, the homomorphism $f^\#: S(X, A) \longrightarrow S(Y, B)$ induces a homomorphism $f_*: H_q(X, A) \longrightarrow H_q(Y, B)$ for all q. It follows immediately from the definition of $f^\#$ that $f^\#$ is the identity when f is. For $f:(X, A) \longrightarrow (Y, B)$ and $g:(Y, B) \longrightarrow (Z, C)$, the definition yields $(gf)^\# = g^\# f^\#$. Lemma 1.2 now yields

4.7. $(gf)_* = g_* f_*$, and $f = \text{Id}$ implies $f_* = \text{Id}$.

Let (X, A, B) be a space triple with $B \subset A \subset X$. The identity map $\text{Id}: X \longrightarrow X$ induces the continuous maps

$$i:(A, B) \longrightarrow (X, B) \text{ and } j:(X, B) \longrightarrow (X, A).$$

4.8. THEOREM: *The sequence*

$$0 \longrightarrow S_q(A, B) \xrightarrow{i^\#} S_q(X, B) \xrightarrow{j^\#} S_q(X, A) \longrightarrow 0$$

is exact.

Proof: Four cases will be considered:

(1) $X = \varnothing$: Then $S_q(\;,\;) = 0$, for all q and all three entries in the parentheses. Thus, Theorem 4.8 is correct in this case.

(2) $X \neq \varnothing$ and $A = B = \varnothing$: Then $S_q(A, B) = 0$ and $i^\# = 0$ for all q. Furthermore, $S_q(X) = S_q(X, A) = S_q(X, B)$ and $j^\# = \text{Id}$. Hence, the sequence is exact.

(3) $X \neq \varnothing$, $A \neq \varnothing$, and $B = \varnothing$: When $q \geq 0$, $gc = \text{Id}c = c$ is to be introduced in the definition of $f^\#$. Then $i_q^\#$ is the injection and

$j_q^{\#}$ is the canonical map on the factor module, and the sequence of Theorem 4.8 is exact. The exactness is trivial here for $q < -1$ and, in the non-augmented case, for $q \leq -1$, since all S_q are zero then. In the augmented case $i_{-1}^{\#}:R(.) \longrightarrow R(.)$ is the identity and $j_{-1}^{\#}:R(.) \longrightarrow 0$, so that exactness also holds here. The case $B = \varnothing$ is precisely the previously treated case of a space pair where $0 \longrightarrow S_q(A) \longrightarrow S_q(X) \longrightarrow S_q(X, A) \longrightarrow 0$ is exact.

(4) $B \neq \varnothing$: When $q < 0$, the sequence is exact since all the modules are zero. There remains the principal case, in which $B \neq \varnothing$ and $q \geq 0$. Here, the exactness of

$$0 \longrightarrow S_q(A)/S_q(B) \xrightarrow{i^{\#}} S_q(X)/S_q(b) \xrightarrow{j^{\#}} S_q(X)/S_q(A) \longrightarrow 0$$

is under question. In this sequence,

$$i^{\#}(c + S_q(B)) = c + S_q(B) \in S_q(X, B) \text{ for } c \in S_q(A),$$

and

$$j^{\#}(d + S_q(B)) = d + S_q(A) \text{ for } d \in S_q(X).$$

Thus, $i^{\#}$ is an injection and $j^{\#}$ is an epimorphism. For $c \in S_q(A)$,

$$j^{\#}i^{\#}(c + S_q(B)) = c + S_q(A) = S_q(A)$$

and thus, $j^{\#}i^{\#} = 0$. If $d \in S_q(X)$ and the element $j^{\#}(d + S_q(B)) = d + S_q(B)$ is the zero of $S_q(X, A)$, then $d \in S_q(A)$, and hence, $d + S_q(B) \in S_q(A, B)$. This implies $i^{\#}(d + S_q(B)) = d + S_q(B)$; in other words, $\text{Ker}(j^{\#}) \subset \text{Im}(i^{\#})$.

4.9. Corollary: *The sequence:*

$$\cdots \longrightarrow H_{q+1}(X, A) \xrightarrow{\partial_*} H_q(A, B) \xrightarrow{i_*} H_q(X, B) \xrightarrow{j_*} H_q(X, A) \xrightarrow{\partial_*} H_{q-1}(A, B) \longrightarrow \cdots$$

is exact by Theorem 1.3.

4.10. THEOREM: *Homeomorphic pairs of spaces have isomorphic homology groups. More precisely: If $f:(X, A) \longrightarrow (Y, B)$ is a homeomorphism; that is, if f induces a homeomorphism f of X on Y such that $f(A) = B$, then f induces an isomorphism of $H_q(X, A)$ on $H_q(Y, B)$.*

Proof: The map $g = f^{-1}$ is a homeomorphism $g:(Y, B) \longrightarrow (X, A)$ such that $gf = \text{Id}$ and $fg = \text{Id}$. Then, by 4.7, $g_* f_* = \text{Id}$ and $f_* g_* = \text{Id}$. Hence, f_* and g_* are isomorphisms.

4.11. THEOREM: *Let $B \subset A \subset X$, $B' \subset A' \subset X'$ and let $f: (X, A, B) \longrightarrow (X', A', B')$ be continuous. Then f induces maps $g: (A, B) \longrightarrow (A', B')$, $h: (X, B) \longrightarrow (X', B')$, and $k: (X, A) \longrightarrow (X', A')$ such that the following diagram is commutative:*

$$\begin{array}{ccccccccc} 0 & \longrightarrow & S_q(A, B) & \xrightarrow{i^\#} & S_q(X, B) & \xrightarrow{j^\#} & S_q(X, A) & \longrightarrow & 0 \\ & & \downarrow g^\# & & \downarrow h^\# & & \downarrow k^\# & & \\ 0 & \longrightarrow & S_q(A', B') & \xrightarrow{i'^\#} & S_q(X', B') & \xrightarrow{j'^\#} & S_q(X', A') & \longrightarrow & 0. \end{array}$$

Proof: The theorem is trivial for the several cases: $X = \varnothing$; $q < -1$; non-augmented complexes when $q = -1$; $q = -1$ and $B \neq \varnothing$; and $q = -1$ and $B' = \varnothing$, since in these cases the upper or the lower row of modules are all zero. In the augmented case when $q = -1$, $X \neq \varnothing$, $B = B' = \varnothing$, and $A \neq \varnothing$, it follows that $A' \neq \varnothing$, and that the same sequence

$$0 \longrightarrow R(.) \longrightarrow R(.) \longrightarrow 0 \longrightarrow 0$$

occupies both the upper and lower rows. Furthermore, since $g^\#$, $h^\#$, and $k^\#$ reduce to the identity, the commutativity is trivial. For $q = -1$, there remains the case for augmented complexes where $X \neq \varnothing$ and $A = B = B' = \varnothing$. When $A' = \varnothing$, the same sequence $0 \longrightarrow 0 \longrightarrow R(.) \longrightarrow R(.) \longrightarrow 0$ occupies both rows and commutativity is trivial. When $A' \neq \varnothing$, the terms $S_{-1}(A, B)$ and $S_{-1}(X', A')$ are both zero and the diagram commutes. There remains the principal case: $q \geq 0$ and $X \neq \varnothing$. Here, $Y \neq \varnothing$.

(a) Let $c \in S_q(X)$. Then

$$h^\# i^\#(c + S_q(B)) = h^\#(c + S_q(B)) = fc + S_q(B'),$$
$$i'^\# g^\#(c + S_q(B)) = i'^\#(fc + S_q(B')) = fc + S_q(B'),$$

where, to avoid the introduction of further notation, the letter f is also used to denote the map of X in X'.

(b) Let $c \in S_q(X)$, then

$$k^\# j^\#(c + S_q(B)) = k^\#(c + S_q(A)) = fc + S_q(A'),$$
$$j'^\# h^\#(c + S_q(B)) = j'^\#(fc + S_q(B')) = fc + S_q(A').$$

4.12. Corollary: *Under the hypotheses of Theorem* 4.11, *each square in*

$$\begin{array}{ccccccccc}
\xrightarrow{\partial_*} & H_q(A, B) & \xrightarrow{i_*} & H_q(X, B) & \xrightarrow{j_*} & H_q(X, A) & \xrightarrow{\partial_*} & H_{q-1}(A, B) & \longrightarrow \cdots \\
 & \downarrow g_* & & \downarrow h_* & & \downarrow k_* & & \downarrow g_* & \\
\xrightarrow{\partial_*} & H_q(A', B') & \xrightarrow{i'_*} & H_q(Y, B') & \xrightarrow{j^*} & H_q(Y, A') & \xrightarrow{\partial_*} & H_{q-1}(A', B') & \longrightarrow \cdots
\end{array}$$

is commutative.

This follows immediately from Theorem 1.4.

Later, in axiomatic study of homology, Theorem 4.11 will be taken as an axiom.

5

Homotopy Properties of Homology Groups

The affine case will be considered first; and, as in Chapter 3, the proofs will be carried out for reduced homology. Let E and F be convex subsets of affine spaces. The set of all affine maps of E in F was denoted by $A_0(E, F)$, and $A(E, F)$ was the free module generated by $A_0(E, F)$. For $q < -1$, $A(\Delta_q, E) = 0$ by definition. In Chapter 3, the notation $x^\rho = \rho(x)$ was introduced for the image of $x \in E$ under the affine map $\rho: E \longrightarrow F$. A homomorphism $\rho: A(\Delta_q, E) \longrightarrow A(\Delta_q, F)$ was induced by $\rho: E \longrightarrow F$. If $q \geq 0$ and $\sigma = (a_0, \cdots, a_q) \in A_0(\Delta_q, E)$ is an affine simplex, then, by definition,

$$\rho(a_0, \cdots, a_q) = (a_0^\rho, \cdots, a_q^\rho) \in A_0(\Delta_q, F).$$

For the elements $c = \sum r_\sigma \sigma$ of $A(\Delta_q, E)$ it follows that $\rho(c) =$

$\sum r_\sigma \rho(\sigma)$. For $q \leq -1$, $\rho : A(\Delta_q, E) \longrightarrow A(\Delta_q, F)$ is defined as the identity.

In addition, it was proved in Chapter 3 that, for $q \geq -1$, each $x \in E$ could be interpreted as a homomorphism $x : A(\Delta_q, E) \longrightarrow A(\Delta_{q+1}, E)$. It was for this purpose that the definition $x(a_0, \cdots, a_q) = (x, a_0, \cdots, a_q)$ was made for $q \geq 0$ and for each affine simplex $(a_0, \cdots, a_q) \in A_0(\Delta_q, E)$. For $q = -1$, $x(.) = (x)$, by definition. It was then easy to see that $\rho x = x^\rho \rho$.

Let λ, μ be two affine maps of E in F. With the resulting homomorphisms $\lambda, \mu : A(\Delta_q, E) \longrightarrow A(\Delta_q, F)$, a recursive homomorphism

$$\psi_q : A(\Delta_q, E) \longrightarrow A(\Delta_{q+1}, F)$$

is defined by means of

5.0. $\psi_q = 0$ for $q \leq -1$

5.1. $\psi_{q+1} x + x^\lambda \psi_q = x^\lambda x^\mu \mu = x^\lambda \mu x$ for $q \geq -1$.

The formulas are meaningful since the homomorphisms of 5.1 are defined on $A(\Delta_q, E)$ and have images in $A(\Delta_{q+2}, F)$. Indeed, the diagram

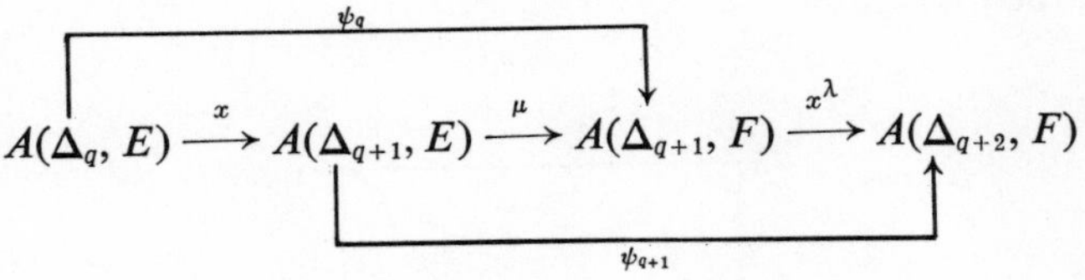

describes the situation.

The map ψ_q can be found explicitly as follows: For $q \leq -1$, $\psi_q = 0$ by 5.0. For $q = -1$, formula 5.1 yields $\psi_0 x + x^\lambda 0 = x^\lambda x^\mu \mu$. By means of $A(\Delta_{-1}, E) = R(.)$, it follows that $\psi_0 x(.) = x^\lambda x^\mu \mu(.)$ or $\psi_0(x) = x^\lambda x^\mu(.) = (x^\lambda, x^\mu)$. Hence, ψ_0 is given by $\psi_0(a_0) = (a_0^\lambda, a_0^\mu)$. Now suppose that ψ_{q-1} has already been defined for $q \geq 1$. Then, for the affine q-simplex $(a_0, \cdots, a_q)$, formula 5.1 yields

$$\begin{aligned} \psi_q(a_0, \cdots, a_q) &= \psi_q a_0(a_1, \cdots, a_q) \\ &= a_0^\lambda a_0^\mu \mu(a_1, \cdots, a_q) - a_0^\lambda \psi_{q-1}(a_1, \cdots, a_q) \\ &= (a_0^\lambda, a_0^\mu, a_1^\mu, a_2^\mu, \cdots, a_q^\mu) - a_0^\lambda \psi_{q-1}(a_1, a_2, \cdots, a_q) \end{aligned}$$

It will now be proved that

$$\textbf{5.2.}\qquad \psi_q(a_0, a_1, \cdots, a_q) = \sum_{i=0}^{q} (-1)^i (a_0^\lambda, \cdots, a_i^\lambda, a_i^\mu, \cdots, a_q^\mu)$$

for $q \geq 0$.

Proof: (By induction.) For $q = 0$, formula 5.2 reduces to $\psi_0(a_0) = (a_0^\lambda, a_0^\mu)$. This has already been proved. Suppose that $q \geq 1$ and that the formula has already been proved for $q - 1$. Then it follows that

$$\begin{aligned}\psi_q(a_0, a_1, \cdots, a_q) &= (a_0^\lambda, a_0^\mu, a_1^\mu, \cdots, a_q^\mu) - a_0^\lambda \psi_{q-1}(a_1, a_2, \cdots, a_q) \\ &= (a_0^\lambda, a_0^\mu, \cdots, a_q^\mu) - a_0^\lambda \sum_{i=1}^{q} (-1)^{i-1} (a_1^\lambda, \cdots, a_i^\lambda, a_i^\mu, \cdots, a_q^\mu) \\ &= \sum_{i=0}^{q} (-1)^i (a_0^\lambda, \cdots, a_i^\lambda, a_i^\mu, \cdots, a_q^\mu).\end{aligned}$$

Geometrical Interpretation for Small q: For $q = 0$, $\psi_0(a_0) = (a_0^\lambda, a_0^\mu)$ is the line segment connecting a_0^λ with a_0^μ. For $q = 1$,

$$\psi_1(a_0, a_1) = (a_0^\lambda, a_0^\mu, a_1^\mu) - (a_0^\lambda, a_1^\lambda, a_1^\mu).$$

Thus, the diagram is

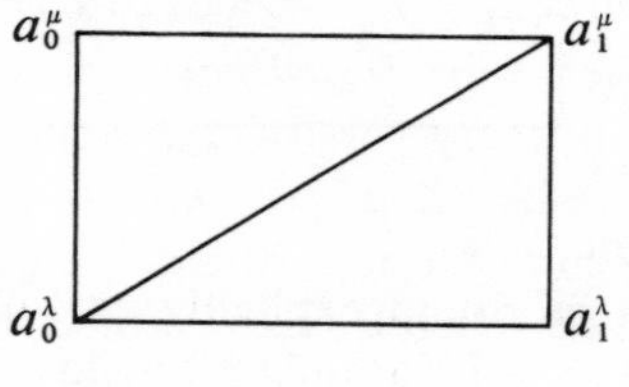

Figure 4

For $q = 2$,

$$\psi_2(a_0, a_1, a_2) = (a_0^\lambda, a_0^\mu, a_1^\mu, a_2^\mu) - (a_0^\lambda, a_1^\lambda, a_1^\mu, a_2^\mu) + (a_0^\lambda, a_1^\lambda, a_2^\lambda, a_2^\mu),$$

which can be illustrated by

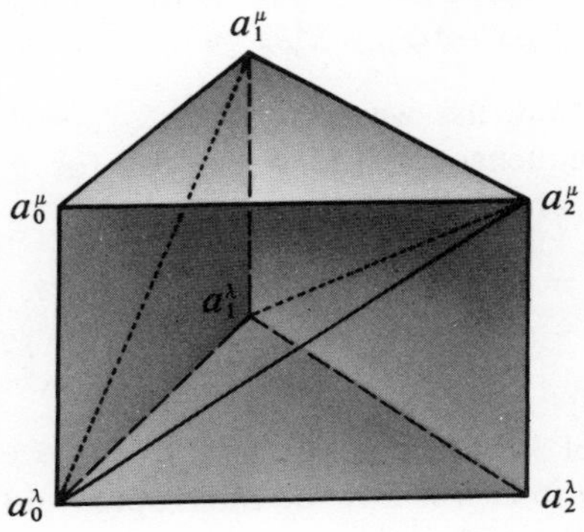

Figure 5

This geometrical interpretation assumes that the points a_0^λ, a_1^λ, a_2^λ, a_0^μ, a_1^μ, a_2^μ are distinct. It is given only to strengthen the intuition and will not be used explicitly. Even when $q \geq 2$, the construction yields a triangulation of the prism with signs on the simplices whenever the vertices $a_0^\lambda, \cdots, a_q^\lambda$, and $a_0^\mu, \cdots, a_q^\mu$ are located in a "reasonable" manner.

5.3. Lemma: *For the boundary of a subdivided prism,*

$$\partial_{q+1}\psi_q + \psi_{q-1}\partial_q = \mu - \lambda \quad \textit{for all } q.$$

Remark on 5.3: Let σ be a q-simplex. Then $\partial_{q+1}\ \psi_q(\sigma)$ is the boundary of the simplicially subdivided prism on σ. According to formula 5.3, this boundary is $\mu(\sigma) - \lambda(\sigma) - \psi_{q-1}\partial_q(\sigma)$. Here (with signs disregarded) $\mu(\sigma)$ is the top, $\lambda(\sigma)$ the base, and $\psi_{q-1}\partial_q(\sigma)$ the simplicially subdivided lateral faces. Thus, 5.3 says that inner boundaries vanish.

Proof of 5.3: For $q \leq -1$, both μ and λ are defined to be the identity, and therefore, $\mu - \lambda = 0$. Since $\psi_q = \psi_{q-1}$ in this case, 5.3 follows for $q \leq -1$.

If ψ_0 is applied to the simplex (x), where $x \in E$, then $\psi_0(x) = (x^\lambda, x^\mu)$ and $\partial_1\psi_0(x) = (x^\mu) - (x^\lambda) = \mu(x) - \lambda(x)$. Since $\psi_{-1} = 0$, 5.3 holds for $q = 0$.

Suppose that 5.3 holds for some $q \geq 0$; then 3.5 and 5.1 yield

$$\begin{aligned}\partial_{q+2}\psi_{q+1}x &= \partial_{q+2}x^\lambda(\mu x - \psi_q) = (\mathrm{Id} - x^\lambda\partial_{q+1})(\mu x - \psi_q)\\ &= \mu x - \psi_q - x_\mu^\lambda\partial_{q+1}x + x^\lambda\partial_{q+1}\psi_q.\end{aligned}$$

In the last equation, use was made of $\mu\partial_{q+1} = \partial_{q+1}\mu$, which is an immediate consequence of 3.6. From 5.3 for q, from 5.1, with $q - 1$ used for q, and from 3.5, it follows that

$$\begin{aligned}\partial_{q+2}\psi_{q+1}x - \mu x &= -\psi_q - x^\lambda\mu\partial_{q+1}x + x^\lambda(\mu - \lambda - \psi_{q-1}\partial_q)\\ &= -\psi_q - x^\lambda\mu\partial_{q+1}x + x^\lambda\mu - \lambda x + (\psi_q x - x^\lambda\mu x)\partial_q)\\ &= -\lambda x - \psi_q(\mathrm{Id} - x\partial q) = -\lambda x - \psi_q\partial_{q+1}x.\end{aligned}$$

An application of 3.3 yields 5.3 for $q + 1$.

The computations will now be interrupted until after Theorem 5.5 so that an indication of their purpose may be given.

Let I denote the interval $0 \leq t \leq 1$, and let $A \subset X$ and $B \subset Y$ be topological spaces.

Definition: Two continuous maps $f, g:(X, A) \longrightarrow (Y, B)$ are called *homotopic* if and only if there is a continuous map $F(x, t)$: $(X \times I, A \times I) \longrightarrow (Y, B)$ such that $F(x, 0) = f(x)$ and $F(x, 1) = g(x)$ for all $x \in X$. This is written $f \sim g$. Intuitively, it means that the map f can be deformed continuously into the map g. The maps $\lambda, \mu:(X, A) \longrightarrow (X \times I, A \times I)$ which are defined by $\lambda(x) = (x, 0)$ and $\mu(x) = (x, 1)$ are homotopic since the identity $G(x, t)$: $(X \times I, A \times I) \longrightarrow (X \times I, A \times I)$ is continuous and satisfies both $G(x, 0) = \lambda(x)$ and $G(x, 1) = \mu(x)$. Some steps will now be taken toward proving that if $f, g:(X, A) \longrightarrow (Y, B)$ are homotopic, then $f_* = g_*$. This will be completed later, in 5.14. If F defines a homotopy between f and g, then $f = F\lambda$ and $g = F\mu$. From $\lambda_* = \mu_*$, it follows from 4.7 that

$$f_* = (F\lambda)_* = F_*\lambda_* = F_*\mu_* = g_*.$$

Thus, it is sufficient to prove that $\lambda_* = \mu_*$. A step in that direction is

5.4. THEOREM: *Given two chain complexes A, B and a chain homomorphism $f:A \longrightarrow B$. If there are homomorphisms $D_q:A_q \longrightarrow B_q$ for which*

5.4a. $$\partial_{q+1}D_q + D_{q-1}\partial_q = f_g,$$

then $f_* = 0$.

Proof: The situation discussed in this theorem is described in the diagram

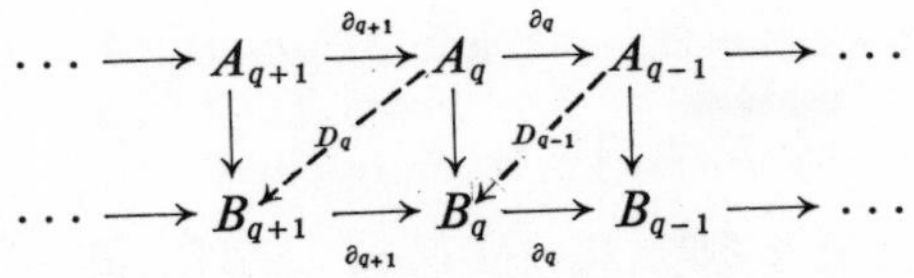

Let $c \in H_q(A)$; that is, $c = a + \partial_{q+1}(A_{q+1})$, where a is a cycle in A_q. By $\partial a = 0$ and 5.4a, it follows that $\partial_{q+1}D_q(a) = f_q(a)$, and hence, that $f_q(a) \in \partial B_{q+1}$. Therefore,

$$f_*(c) = f_q(a) + \partial_{q+1}(B_{q+1}) = \partial_{q+1}B_{q+1},$$

where $\partial_{q+1} B_{q+1}$ is the zero element of $H_q(B)$. Thus, $f_* = 0$.

Comparison of 5.3 and 5.4a shows that preparations are under way for the construction of a D_q which will suffice to prove $(\mu - \lambda)_* = \mu_* - \lambda_* = 0$.

Definition: Two chain homomorphisms $\alpha, \beta: A \longrightarrow B$ are called *chain homotopic* (written $\alpha \sim \beta$) if and only if there are homomorphisms D_q for which $\partial_{q+1}D_q + D_{q-1}\partial_q = \alpha_q - \beta_q$. From $\alpha \sim \beta$, it follows by Theorem 5.4 that $\alpha_* = \beta_*$.

5.5. THEOREM: *The relation $\sim$ is an equivalence relation. When, for $\alpha, \alpha': A \longrightarrow B$ and $\beta, \beta': A \longrightarrow B$, the relations $\alpha \sim \alpha'$ and $\beta \sim \beta'$ hold, then $r\alpha + s\beta \sim r\alpha' + s\beta'$ for $r, s \in R$. When, for $\alpha, \alpha': A \longrightarrow B$ and $\beta, \beta': B \longrightarrow C$, the relations $\alpha \sim \alpha'$ and $\beta \sim \beta'$ hold, then $\beta\alpha \sim \beta'\alpha'$.*

Proof: The relation $\alpha \sim \alpha$ is obtainable by the choice $D_q = 0$, for all q. If $\alpha \sim \alpha'$ and D_q is replaced by $-D_q$, then $\alpha' \sim \alpha$. If

$$\partial D_q + D_{q-1}\partial = \alpha_q - \alpha'_q$$

and

$$\partial D'_q + D'_{q-1}\partial = a'_q - a''_q,$$

then use of $D_q + D'_q$ shows that $\alpha \sim \alpha'$. Thus, $\sim$ is an equivalence relation. If $\alpha, \alpha', \beta, \beta': A \longrightarrow B$, and if

$$\partial D_q + D_{q-1}\partial = \alpha_q - \alpha'_q \text{ and } \partial D'_q + D'_{q-1}\partial = \beta_q - \beta'_q,$$

then $rD_q + sD'_q$ yields the desired result. For the final conclusion of 5.5, suppose that

$$\partial D_q + D_{q-1}\partial = \alpha_q - \alpha'_q \text{ and } \partial D'_q + D'_{q-1}\partial = \beta - \beta'.$$

Let $\bar{D}_q = \beta'_{q+1}D_q + D'_q\alpha_q$. Then,

$$\begin{aligned}\partial\bar{D}_q + \bar{D}_{q+1}\partial &= \partial\beta'_{q+1}D_q + \partial D'_q\alpha_q + \beta'_q D_{q-1}\partial + D'_{q-1}\alpha_{q-1}\partial \\ &= \beta'_q(\partial D_q + D_{q-1}\partial) + (\partial D'_q + D'_{q-1}\partial)\alpha_q \\ &= \beta'_q(\alpha_q - \alpha'_q) + (\beta_q - \beta'_q)\alpha_q \\ &= \beta_q\alpha_q - \beta'_q\alpha'_q\end{aligned}$$

because $\partial\beta'_{q+1} = \beta'_q\partial$ and $\partial\alpha_q = \alpha_{q-1}\partial$ since α and β' are chain homomorphisms.

In the following computation, an abbreviated notation will be used: Let X and Y be topological spaces and let $\phi: X \longrightarrow Y$ be a continuous map. Then $\tilde{\phi}: X \times I \longrightarrow Y \times I$ shall denote the map which is defined by $\tilde{\phi}((x, t)) = (\phi(x), t)$ or by $(x, t)^{\tilde{\phi}} = (x^{\phi}, t)$. The continuity of ϕ implies continuity of $\tilde{\phi}$. When $X \xrightarrow{\phi} Y \xrightarrow{\theta} Z$, it is easy to verify that $\widetilde{(\theta\phi)} = \tilde{\theta}\tilde{\phi}$. The association $\phi \longrightarrow \tilde{\phi}$ and a linear extension yields a monomorphism of $C(X, Y)$ into $C(X \times I, Y \times I)$. If X and Y are affine spaces and ϕ is affine, then $X \times I$ and $Y \times I$ are affine spaces and $\tilde{\phi}$ is affine. A monomorphism of $A(X, Y)$ into $A(X \times I, Y \times I)$ is therefore induced.

Now let E be a convex subset of an affine space. Then $E \times I$ is also convex. Let $q \geq 0$ and let $\sigma = (a_0, a_1, \cdots, a_q)$ be an affine simplex of E; in other words, $\sigma: \Delta_q \longrightarrow E$. Then $\tilde{\sigma}: \Delta_q \times I \longrightarrow E \times I$ is an affine map of the unsubdivided prism $\Delta_q \times I$ into $E \times I$. Now define

$$\lambda: E \longrightarrow E \times I \text{ by means of } \lambda(x) = x^{\lambda} = (x, 0)$$

5.6.

$$\mu: E \longrightarrow E \times I \text{ by means of } \mu(x) = x^{\mu} = (x, 1).$$

Equations 5.0 and 5.1 yield functions $\psi_q: A(\Delta_q, E) \longrightarrow A(\Delta_{q+1}, E \times I)$ such that $\psi_{q+1}x + x^{\lambda}\psi_q = x^{\lambda}x^{\mu}\mu$, for $q \geq -1$, and such that $\psi_{-1} = 0$.

Suppose that E has been specialized to Δ_q and that $\Delta_q^{\dagger}$, as before, is the identity map of Δ_q on Δ_q. Then,

$$\psi_q(\Delta_q^\dagger) \in A(\Delta_{q+1}, \Delta_q \times I).$$

Composition with the element $\tilde{\sigma} \in A(\Delta_q \times I, E \times I)$ yields $\tilde{\sigma}\psi_q(\Delta_q^\dagger) \in A(\Delta_{q+1}, E \times I)$ for which

5.7. $$\tilde{\sigma}\psi_q(\Delta_q^\dagger) = \psi_q\sigma \quad \text{when } q \geq 0.$$

Proof: By 5.2,

$$\tilde{\sigma}\psi_q(\Delta_q^\dagger) = \sum_{i=1}^{q} (-1)^i \tilde{\sigma}(d_0^\lambda, \cdots, d_i^\lambda, d_i^\mu, \cdots, d_q^\mu).$$

By definition, $\tilde{\sigma}(x, t) = (x^\sigma, t)$ for $x \in \Delta_q$; therefore,

$$\tilde{\sigma}(d_j^\lambda) = \tilde{\sigma}(d_j, 0) = (d_j^\sigma, 0) = (a_j, 0) = a_j^\lambda.$$

Analogous formulas hold for μ. Consequently,

$$\tilde{\sigma}\psi_q(\Delta_q^\dagger) = \sum_{i=0}^{q} (-1)^i (a_0^\lambda, \cdots, a_i^\lambda, a_i^\mu, \cdots, a_q^\mu) = \psi_q(\sigma).$$

Equation 5.7 extends readily to hold for linear combinations of affine simplices σ and thus applies to all $c \in A(\Delta_q, E)$.

An application of 5.3 to $\Delta_q^\dagger$ yields

$$\partial_{q+1}\psi_q(\Delta_q^\dagger) + \psi_{q-1}\partial_q(\Delta_q^\dagger) = \mu(\Delta_q^\dagger) - \lambda(\Delta_q^\dagger).$$

By 3.9, $\partial_q c = c\partial_q(\Delta_q^\dagger)$ for $c \in A(\Delta_q, E)$, and then,

$$\partial_{q+1}\psi_q(\Delta_q^\dagger) = \psi_q(\Delta_q^\dagger)\partial_{q+1}(\Delta_q^\dagger).$$

The second term, $\psi_{q-1}\partial_q(\Delta_q^\dagger)$, can be transformed by 5.7, when $q \geq 1$ and $\sigma = \partial_q(\Delta_q^\dagger)$, to yield

$$\psi_{q-1}\partial_q(\Delta_q^\dagger) = \widetilde{(\partial_q(\Delta_q^\dagger))}\psi_{q-1}(\Delta_{q-1}^\dagger).$$

For $q = 0$, the second term is zero since $\psi_{-1} = 0$. This computation yields

5.8a. $$\begin{aligned} \psi_q(\Delta_q^\dagger)\partial_{q+1}(\Delta_{q+1}^\dagger) + \widetilde{(\partial_q(\Delta_q^\dagger))}\psi_{q-1}(\Delta_{q-1}^\dagger) &= \partial_{q+1}\psi_q(\Delta_q^\dagger) + \psi_{q-1}\partial_q(\Delta_q^\dagger) \\ &= \mu(\Delta_q^\dagger) - \lambda(\Delta_q^\dagger), \text{ when } \quad q \geq 1 \end{aligned}$$

and

5.8b. $$\psi_0(\Delta_0^\dagger)\partial_1(\Delta_1^\dagger) = \mu(\Delta_0^\dagger) - \lambda(\Delta_0^\dagger).$$

This completes the computation of $\mu - \lambda$ in the affine case.

Now let (X, A) be a pair of topological spaces; i.e., $A \subset X$. When $q \geq 0$ and $c \in C(\Delta_q, X)$, then

$$\tilde{c} \in C(\Delta_q \times I, X \times I) \text{ and } \psi_q(\Delta_q^\dagger) \in C(\Delta_{q+1}, \Delta_q \times I).$$

It is therefore possible to form the product $\tilde{c}\psi_q(\Delta_q^\dagger) \in C(\Delta_{q+1}, X \times I)$. When $c \in C(\Delta_q, X)$, then $\tilde{c}\psi_q(\Delta_q^\dagger) \in C(\Delta_{q+1}, A \times I)$.

In both the augmented and the non-augmented cases, a homomorphism $\psi_q : S_q(X, A) \longrightarrow S_{q+1}(X \times I, A \times I)$ is now defined for all q by

5.9.
$$\psi_q = 0 \quad \text{for } q < 0,$$
$$\psi_q(c + C(\Delta_q, A)) = \tilde{c}\psi_q(\Delta_q^\dagger) + C(\Delta_{q+1}, A \times I) \quad \text{for } q \geq 0 \text{ and } c \in C(\Delta_q, X).$$

It was just proved that $\psi_q(C(\Delta_q, A)) \subset C(\Delta_{q+1}, A \times I)$. Lemma 1.1 now implies that ψ_q is well defined.

The next computation is that of $\partial_{q+1}\psi_q + \psi_{q-1}\partial_q$ for $q \geq 0$. Equation 3.9 yields $\partial_{q+1}(\tilde{c}\psi_q(\Delta_q^\dagger)) = \tilde{c}\psi_q(\Delta_q^\dagger)\partial_{q+1}(\Delta_{q+1}^\dagger)$. Then from 5.9 it follows that

$$\partial_{q+1}\psi_q(c + C(\Delta_q, A)) = \tilde{c}\psi_q(\Delta_q^\dagger)\partial_{q+1}(\Delta_{q+1}^\dagger) + C(\Delta_q, A \times I).$$

When $q = 0$, relation 5.8b and $\psi_{-1} = 0$ yield

5.10.
$$(\partial_1\psi_0 + \psi_{-1}\partial_0)(c + C(\Delta_0, A)) = \tilde{c}(\mu(\Delta_0^\dagger) - \lambda(\Delta_0^\dagger)) + C(\Delta_0, A \times I).$$

For the second summand, $\psi_{q-1}\partial_q$, Equation 5.7 yields, in case $q \geq 1$,

$$\begin{aligned}\psi_{q-1}\partial_q(c + C(\Delta_q, A)) &= \psi_{q-1}(c\partial_q(\Delta_q^\dagger) + C(\Delta_{q-1}, A)) \\ &= \widetilde{c\partial_q(\Delta_q^\dagger)}\psi_{q-1}(\Delta_{q-1}^\dagger) + C(\Delta_q, A \times I) \\ &= \tilde{c}\widetilde{\partial_q(\Delta_q^\dagger)}\psi_{q-1}(\Delta_{q-1}^\dagger) + C(\Delta_q, A \times I).\end{aligned}$$

Use of 5.8a finally yields

5.11.
$$(\partial_{q+1}\psi_q + \psi_{q-1}\partial_q)(c + C(\Delta_q, A)) = \tilde{c}(\mu(\Delta_q^\dagger) - \lambda(\Delta_q^\dagger)) + C(\Delta_q, A \times I).$$

Because of 5.10, this equation holds for $q \geq 0$. The term $\tilde{c}\mu(\Delta_q^\dagger)$

will now be transformed: For a q-simplex $\sigma: \Delta_q \longrightarrow X$, the function $\tilde{\sigma}:\Delta_q \times I \longrightarrow X \times I$ is given by $\tilde{\sigma}(y, t) = (\sigma(y), t)$. By definition, $\mu(\Delta_q^\dagger) = (d_0^\mu, d_1^\mu, \cdots, d_q^\mu)$; then, $\mu(\Delta_q^\dagger)y = (y, 1) \in \Delta_q \times I$ for $y \in \Delta_q$. Hence, $\tilde{\sigma}\mu(\Delta_q^\dagger)y = (\sigma(y), 1)$. If the function $\mu':X \longrightarrow X \times I$ is temporarily defined by $\mu'(x) = (x, 1)$, then it follows that $\mu'\sigma(y) = (\sigma(y), 1)$. Consequently, $\tilde{\sigma}\mu(\Delta_q^\dagger) = \mu'\sigma$, and hence, $\tilde{c}\mu(\Delta_q^\dagger) = \mu'c$. Similarly, $\tilde{c}\lambda(\Delta_q^\dagger) = \lambda'c$ provided λ' is defined by $\lambda'(x) = (x, 0)$ for $x \in X$. Therefore, for $q \geq 0$ and $c \in C(\Delta_q, X)$, it follows that

$$(\partial_{q+1}\psi_q + \psi_{q-1}\partial_q)(c + C(\Delta_q, A)) = (\mu' - \lambda')c + C(\Delta_q, A \times I).$$

The reader may now forget the intermediate computations and omit the primes from μ' and λ'. This means that to the functions $\mu:X \longrightarrow X \times I$ and $\lambda:X \longrightarrow X \times I$ defined by $\mu(x) = (x, 1)$ and $\lambda(x) = (x, 0)$, respectively, there have been found continuous functions $\psi_q:S_q(X, A) \longrightarrow S_{q+1}(X \times I, A \times I)$ such that

5.12. $$(\partial_{q+1}\psi_q + \psi_{q-1}\partial_q)(c + C(\Delta_q, A)) = (\mu - \lambda)c + C(\Delta_q, A \times I) \qquad \text{for } q \geq 0.$$

The explicit meaning of ψ_q is no longer of any particular interest. The maps λ and μ induce chain homomorphisms

$$\lambda_q^\#, \mu_q^\#: S_q(X, A) \longrightarrow S_q(X \times I, A \times I)$$

where $\mu_q^\#(c + S_q(A)) = \mu c + S_q(A \times I)$ for $q \geq 0$ and $X \neq \varnothing$. Then 5.12 implies that

5.13. $$\partial_{q+1}\psi_q + \psi_{q-1}\partial_q = \mu_q^\# - \lambda_q^\# \qquad \text{for } q \geq 0.$$

Equation 5.13 also holds when $q < 0$, since the left side is 0 and on the right either $\mu_q^\# = \lambda_q^\# = 0$, or else (in the augmented case when $q = -1$ and $A = \varnothing$) $\mu_q^\# = \lambda_q^\# = Id$. Now Theorem 5.4 can be applied, with ψ_q in place of D_q and $\mu^\# - \lambda^\#$ in place of f, to yield $\mu_* = \lambda_*$. This proves the *Homotopy Theorem* (5.14):

5.14. THEOREM: *Let $\lambda, \mu:(X, A) \longrightarrow (X \times I, A \times I)$ be given by $\lambda(x) = (x, 0)$ and $\mu(x) = (x, 1)$. Then the induced maps $\lambda_*, \mu_*:H_q(X, A) \longrightarrow H_q(X \times I, A \times I)$ are equal;* i.e. $\mu_* = \lambda_*$.

5.15. Corollary: *If $f,g:(X, A) \longrightarrow (Y, B)$ are homotopic, then $f_* = g_*$.*

Proof: The proof was carried out just prior to 5.4.

5.16. Lemma: *Let X and Y be topological spaces and let X_1, X_2 be closed subspaces of X. Let $f_i:X_i \longrightarrow Y$ be continuous for $i = 1, 2$ and let $f_1(x) = f_2(x)$ for $x \in X_1 \cap X_2$. Then the function $f:X_1 \cup X_2 \longrightarrow Y$, defined by $f(x) = f_i(x)$ for $x \in X_i$, is continuous.*

Proof: Let A be a closed set in Y. Then $f_i^{-1}(A)$ is closed in X_i. Since X_i is closed, $f_i^{-1}(A)$ is closed in X. Hence, $f^{-1}(A) = f_1^{-1}(A) \cup f_2^{-1}(A)$ is closed in $X_1 \cup X_2$, and f is continuous.

The lemma and proof remain correct if the word "closed" is replaced by "open" throughout. The hypotheses can be weakened, although this is not of interest here. The next theorem is an application of the lemma.

5.17. THEOREM: *If $f, g, h:(X, A) \longrightarrow (Y, B)$ are continuous and if $f \sim g$ and $g \sim h$, then $f \sim h$.*

Proof: There exist continuous functions $F, G:(X \times I, A \times I) \longrightarrow (Y, B)$ with $F(x, 0) = f(x)$, $F(x, 1) = g(x)$, $G(x, 0) = g(x)$, $G(x, 1) = h(x)$. Let I' be the interval $0 \leq t \leq \frac{1}{2}$ and I'' be the interval $\frac{1}{2} \leq t \leq 1$. Then $X \times I'$ and $X \times I''$ are closed subspaces of $X \times I$ whose intersection is the set $\{(x, \frac{1}{2}) \mid x \in X\}$. The function defined by $F(x, 2t)$ is continuous on $X \times I'$ and that defined by $G(x, 2t - 1)$ is continuous on $X \times I''$. By 5.16, the function H defined by

$$H(x, t) = \begin{cases} F(x, 2t) & \text{for} \quad t \in I' \\ G(x, 2t - 1) & \text{for} \quad t \in I'' \end{cases}$$

is continuous. Furthermore $H(x, 0) = f(x)$, $H(x, 1) = h(x)$. Thus, $f \sim g$.

5.18. Corollary: *The relation $\sim$ is an equivalence relation.*

Proof: The transitivity has just been proved. For reflexivity, define $F(x, t) = f(x)$. For symmetry, use $F(x, 1 - t)$ in place of $F(x, t)$.

5.19. THEOREM: *Let $f_1, f_2:(X, A) \longrightarrow (Y, B)$ and $g_1, g_2: (Y, B) \longrightarrow (Z, C)$ be continuous. Let $f_1 \sim f_2$ and $g_1 \sim g_2$. Then $g_1 f_1 \sim g_2 f_2$.*

Proof: Let $F:(X \times I, A \times I) \longrightarrow (Y, B)$ be a continuous function for which $F(x, 0) = f_1(x)$ and $F(x, 1) = f_2(x)$. If $g:(U, 0) \longrightarrow (X, A)$

and $h:(Y, B) \longrightarrow \mathrm{Im}(h)$ are continuous, then the maps defined by $F(g(u), t)$ and $h(F(x, t))$ are continuous irrespective of $\mathrm{Im}(h)$. Because $F(g(u), 0) = f_1g(u)$, $F(g(u), 1) = f_2g(u)$, $h(F(x, 0)) = hf_1(x)$, and $h(F(x, 1)) = hf_2(x)$, it follows that $f_1g \sim f_2g$ and $hf_1 \sim hf_2$. For the maps of the hypothesis it therefore follows that $g_1f_1 = g_2f_2$ and $g_2f_1 \sim g_2f_2$. Transitivity then yields $g_1f_1 \sim g_2f_2$.

Definition: If, to a continuous function $f:(X, A) \longrightarrow (Y, B)$, there is a continuous function $g:(Y, B) \longrightarrow (X, A)$, such that $gf:(X, A) \longrightarrow (X, A)$ and $fg(Y, B) \longrightarrow (Y, B)$ are both homotopic to the identity, then g is called a *homotopy inverse* of f.

5.20. *The following hold for homotopy inverses:*

(1) *If f has a homotopy inverse g and if $f' \sim f$, then g is also a homotopy inverse for f';* because $f'g \sim fg \sim \mathrm{Id}$ and $gf' \sim gf \sim \mathrm{Id}$.

(2) *If g is a homotopy inverse of f and $g' \sim g$, then g' is also a homotopy inverse for f.* The proof is as in (1).

(3) *If g and g' are homotopy inverses of f, then $g \sim g'$.* This follows from $g \sim g\,\mathrm{Id} \sim gfg' \sim \mathrm{Id}\,g' \sim g$.

Definition: A function $f:(X, A) \longrightarrow (Y, B)$ which has a homotopy inverse is called a *homotopy equivalence*. Two pairs of spaces (X, A) and (Y, B) for which there is a homotopy equivalence $f:(X, A) \longrightarrow (Y, B)$ are called *homotopic.*

5.21. THEOREM: *Let $f_1(X, A) \longrightarrow (Y, B)$ and $f_2:(Y, B) \longrightarrow (Z, C)$ be homotopy equivalences. Then $f_2f_1:(X, A) \longrightarrow (Z, C)$ is a homotopy equivalence.*

Proof: Let g_1 and g_2 be the respective homotopy inverses of f_1 and f_2. Then $f_2f_1g_1g_2 \sim f_2\,\mathrm{Id}\,g_2 \sim f_2g_2 \sim \mathrm{Id}$, and similarly, $g_1g_2f_2f_1 \sim \mathrm{Id}$. Thus, g_1g_2 is a homotopy inverse of f_2f_1.

5.22. Corollary: *Homotopy of space pairs is an equivalence relation.*

Proof: The transitivity was proved in 5.21. Symmetry follows from an interchange of f and g. Reflexivity follows from letting $f = g = \mathrm{Id}$.

5.23. Remark: *Homeomorphic space pairs are homotopic* since f^{-1} is a homotopy inverse of f.

5.24. By the homotopy theorem (5.14) it follows for homotopic maps $f, g\colon (X, A) \longrightarrow (Y, B)$ that $f_* = g_*$. If f has a homotopy inverse g, then it follows that $(fg)_* = \mathrm{Id}$ and $(gf)_* = \mathrm{Id}$. Hence, $f_*\colon H_q(X, A) \longrightarrow H_q(Y, B)$ is an isomorphism. Thus, *homotopic space pairs have isomorphic homology groups.*

Example of a Homotopy Equivalence:
Let $\lambda\colon(X, A) \longrightarrow (X \times I, A \times I)$ be defined by $\lambda(x) = (x, 0)$ and $pr\colon(X \times I, A \times I) \longrightarrow (X, A)$ by $pr(x, t) = x$. That λ is a homotopy equivalence with pr as homotopy inverse can be seen as follows: Clearly, $pr\,\lambda = \mathrm{Id}$. The function

$$F\colon(X \times I \times I, A \times I \times I) \longrightarrow (X \times I, A \times I)$$

which is defined by $F(x, t, s) = (x, ts)$ shows that $\lambda\, pr \sim \mathrm{Id}$ since

$$\lambda\, pr(x, t) = (x, 0) = F(x, t, 0)$$

and

$$\mathrm{Id}(x, t) = (x, t) = F(x, t, 1).$$

5.25. Corollary: *The function* $\lambda\colon(X, A) \longrightarrow (X \times I, A \times I)$ *given by* $\lambda(x) = (x, 0)$ *defines an isomorphism* $\lambda_*\colon H_q(X, A) \longrightarrow H_q(X \times I, A \times I)$.

Geometrical Consequences of the Homotopy Theorem

Definition: Let Y be a subspace of X and r a continuous function $r\colon X \longrightarrow Y$ which is the identity on Y. Then r is a *retraction* of X on Y and Y is a *retract* of X.

Not every subspace is a retract. For example, let X be a closed interval with interior points and let Y be the pair of endpoints of X. Since continuous images of connected sets are connected, a retract of X would need to be connected. Thus, Y cannot be a retract of X, for Y is not connected.

Definition: Let (Y, B) and (X, A) be space pairs with $(Y, B) \subset (X, A)$; that is to say, $Y \subset X$ and $B \subset A$. Let $r\colon(X, A) \longrightarrow (Y, B)$

be a continuous function for which $r(y) = y$ for all y. Then r is called a *retraction* of (X, A) on (Y, B) and (Y, B) is called a *retract* of (X, A).

Let $i:(Y, B) \longrightarrow (X, A)$ denote the injection. Then, for the retraction r, $ri:(Y, B) \longrightarrow (Y, B)$ is the identity. Thus, r is the homotopy inverse of i if and only if $ir:(X, A) \longrightarrow (X, A)$ is homotopic to the identity.

Definition: A retraction r of (X, A) onto (Y, B) is called a *deformation retraction*, and (Y, B) is called a *deformation retract* of (X, A) if and only if ir is homotopic to the identity.

From the Homotopy Theorem (5.14), it follows for a deformation retraction $r:(X, A) \longrightarrow (Y, B)$ that $r_*^{-1} = i_*$ and that $i_*:H_q(Y, B) \longrightarrow H_q(X, A)$ is an isomorphism. (The reader will remember that isomorphisms are epimorphisms.)

A retraction $r:(X, A) \longrightarrow (Y, B)$ is a deformation retraction if and only if there is a continuous function $F:(X \times I, A \times I) \longrightarrow (X, A)$ with $F(x, 0) = x$ and $F(x, 1) = ir(x) = r(x)$. For $y \in Y$, $F(y, 1) = r(y) = y$.

Definition: Let $r:(X, A) \longrightarrow (Y, B)$ be a deformation retraction. Then (Y, B) is a *strong deformation retract* of (X, A) if and only if $F(y, t) = y$ for all $y \in Y$ and $t \in I$.

5.26. Examples:

(1) Let $P \neq Q$ be two points and let $X = P \cup Q$, $Y = P$, $A = \varnothing$. Then Y is a retract of X but not a deformation retract since F maps the connected set $Q \times I$ on a connected set (and hence, a point) and it therefore is impossible that both $F(Q, 0) = Q$ and $F(Q, 1) = r(Q) = P$.

(2) Let X consist of two circles which have a single common point. Let Y be one of the circles and let $A = \varnothing$. Folding shows that Y is a retract of X. It will be proved later that Y is not a deformation retract of X.

(3) *Exercise:* Let X consist of the points (x, y) of R^2 for which either $0 \leq x \leq 1$ and $y = 0$, or $x = 0$ and $0 \leq y \leq 1$, or $x = 1/n$, $0 \leq y \leq 1$, where $n = 1, 2, 3, \cdots$. Let Y be the point $(0, 1)$ and $A = \varnothing$. Then Y is a deformation retract, but not a strong deformation retract of X.

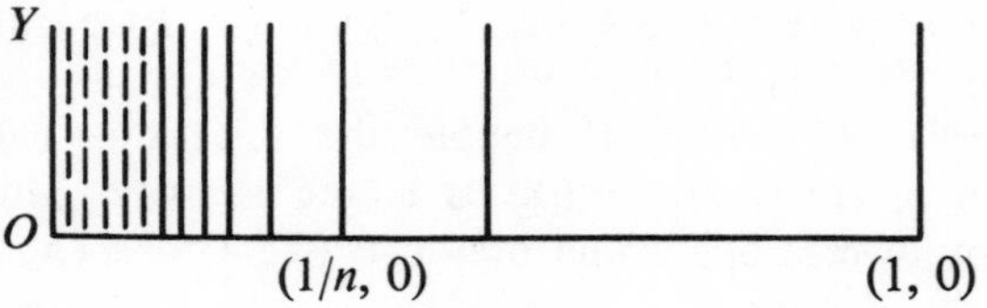

Figure 6

(4) Let X be the ball $|x| \leq r$, that is, the set of all points of R^n with position vector x and $|x| \leq r$. Let $A = \varnothing$ and let Y be the origin. Then Y is a strong deformation retract of X, as can be seen from $F(x, t) = tx$.

(5) *Exercise:* Let X be the solid torus. The central circle is a strong deformation retract of X.

(6) *Exercise:* Let X be a torus (surface) in which a small circular hole has been cut. Describe a pair of circles which have just one common point that form a deformation retract of X.

(7) The contraction of a spike is often employed: Let $X = A \cup S$ be a subset of R^n where $A \cap S =$ origin and $S = \{x \mid x \in R^n, 0 \leq x_1 \leq 1, x_i = 0 \text{ for } i = 2, \cdots, n\}$. Let $F(x, t) = tx$ for $x \in S$ and $F(x, t) = x$ for $x \in A$. Then $F:(X \times I) \longrightarrow X$ is continuous and satisfies $F(x, 1) = x$ for all $x \in X$ and $F(x, 0) = r(x) = x$ for $x \in A$ and $F(x, 0) = r(x) = 0$ for $x \in S$. Thus, A is a deformation retract of $X = A \cup S$.

(8) Let H be a hypersurface in R^n with the following property: To each vector $x \neq 0$ there is precisely one intersection point $h(x)$ of H with the half line $\lambda x, \lambda > 0$, and $h(x)$ is a continuous function of x. Then H is a deformation retract of each of the three sets:

$$X_0 = \{x \mid x \neq 0\}, \ X_1 = \{x \mid |x| \geq |h(x)|\},$$
$$\text{and } X_2 = \{x \mid x \neq 0, \ |x| \leq |h(x)|\}.$$

Proof: The function

$$F(x, t) = \frac{|x| + t\,|h(x)|}{t\,|x| + |h(x)|}\, h(x)$$

is continuous for $x \neq 0$, $0 \leq t \leq 1$. Furthermore,

$$F(x, 0) = \frac{|x|\, h(x)}{|h(x)|} = x$$

$$F(x, 1) = h(x), \text{ and } F(h(x), t) = h(x).$$

Thus, H is a strong deformation retract of X_0. Since the inequality $|x| \leq |h(x)|$ implies that

$$|x| + t|h(x)| \leq t|x| + |h(x)|,$$

which implies that $|F(x, t)| \leq |h(x)|$, the set $X_i \times I$ is mapped onto X_i by $F(x, t)$ for $i = 1, 2$. Hence, H is also a strong deformation retract of both X_1 and X_2.

(9) As an application of (8), consider R^3 with $(-1, 0, 0)$ and $(1, 0, 0)$ removed. By (8), the sphere $|x| = 2$ is a strong deformation retract of $\{x \mid |x| \geq 2\}$. Let the point $(-1, 0, 0)$ be surrounded by a surface H_1 which consists of the hemisphere $|x| = 2$, $x_1 < 0$ and the disc $|x| \leq 2$, $x_1 = 0$. By (8), H_1 is a strong deformation retract of the punctured interior of H_1. If H_2 is defined analogously so that it surrounds $(1, 0, 0)$, then $H_1 \cup H_2$ is a strong deformation retract of the twice-punctured R^3.

Application to Graphs

Definition: A *graph* is a connected union of a finite collection of topological line segments which have at most end points in common. A topological circle is therefore one example of a graph. Graphs can be embedded in R^3, but in many essentially different ways; for instance, the circle and the trefoil knot are homeomorphic. The nature of the embedding of a graph is a difficult question. However, the problem of the homotopy and homology of a graph is easily settled.

(1) A segment with a free end point, a spike in other words, can be eliminated by a deformation retraction upon the other end point as described in 5.26 (7).

(2) Let P, Q with $P \neq Q$ be end points of a segment of the graph T. Let P be an end point of at least three segments. On one of these which is not PQ choose a point R so that R, P, Q become vertices of a triangle which has only the sides RP and PQ in common with the graph. Let T' be the graph that results when one connects R with Q instead of with P. The diagrams show that both T and T' are deformation retracts of the union

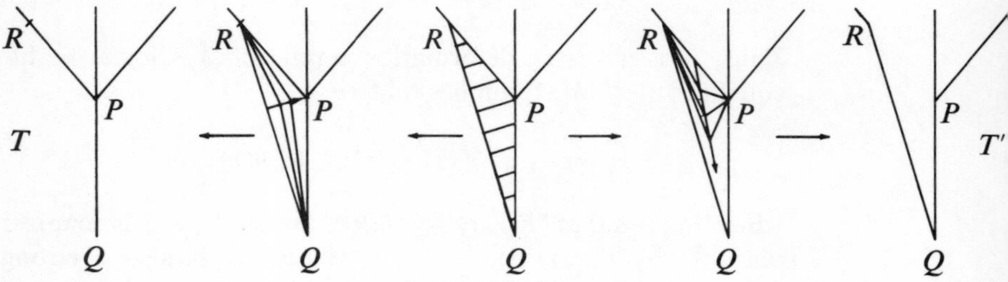

Figure 7

of T with the triangle RPQ. Hence, T is homotopic to T'. In the formation of T', the number of segments leaving P has been reduced by one. By a repetition of this procedure, the number of segments leaving P can be reduced to two. Such a result is equivalent to the contraction of P to Q, and can be repeated until all segments have the same point Q as their beginning and end points.

(3) It was just proved that each graph is homotopic to a union of finitely many topological circles that have exactly one point Q in common.

Figure 8

(4) This procedure also shows that a line segment with the same number of attached circles is homotopic to the figure arrived at in (3).

The number of circles will be proved to be a homotopy invariant. It is also a homology invariant as will be shown by a computation of the homology groups.

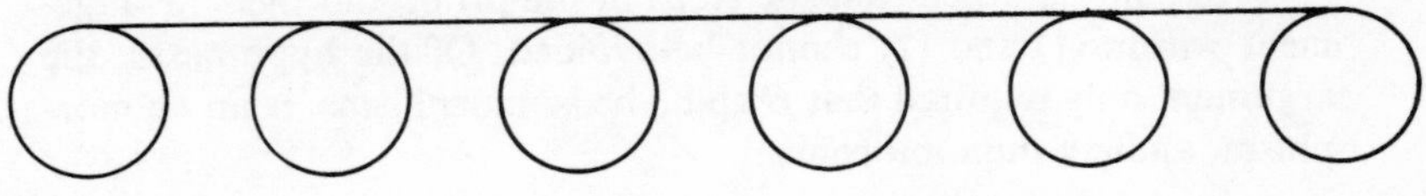

Figure 9

5.27. The "Five" Lemma: *In the following diagram let the rows be exact, and let the rectangles be commutative. Furthermore, let α, β, δ, ϵ be isomorphisms. Then γ is also an isomorphism.*

$$\begin{array}{ccccccccc} A_1 & \xrightarrow{\phi_1} & A_2 & \xrightarrow{\phi_2} & A_3 & \xrightarrow{\phi_3} & A_4 & \xrightarrow{\phi_4} & A_5 \\ \downarrow{\scriptstyle\alpha} & & \downarrow{\scriptstyle\beta} & & \downarrow{\scriptstyle\gamma} & & \downarrow{\scriptstyle\delta} & & \downarrow{\scriptstyle\epsilon} \\ B_1 & \xrightarrow{\psi_1} & B_2 & \xrightarrow{\psi_2} & B_3 & \xrightarrow{\psi_3} & B_4 & \xrightarrow{\psi_4} & B_5. \end{array}$$

Proof:

(1) Let $\gamma(a_3) = 0$. Then $\psi_3\gamma(a_3) = \delta\phi_3(a_3) = 0$, and hence, $\phi_3(a_3) = 0$. Therefore, there is an a_2 such that $a_3 = \phi_2(a_2)$ and

$$\gamma(a_3) = \gamma\phi_2(a_2) = \psi_2\beta(a_2) = 0.$$

Hence, for some a_1,

$$\beta(a_2) = \psi_1(b_1) = \psi_1\alpha(a_1) = \beta\psi_1(a_1).$$

Thus, γ is a monomorphism.

(2) The problem now is to find an inverse element for $b_3 \in B_3$. Since δ is an isomorphism, $\psi_3(b_3) = \delta(a_4)$, and hence,

$$0 = \psi_4\psi_3(b_3) = \psi_4\delta(a_4) = \epsilon\phi_4(a_4).$$

Therefore $\phi_4(a_4) = 0$, and there is an $a_3 \in A_3$ such that $a_4 = \phi_3(a_3)$. Then

$$\psi_3(b_3) = \delta(a_4) = \delta\phi_3(a_3) = \psi_3\gamma(a_3),$$

from which it follows that there is a b_2 such that

$$b_3 - \gamma(a_3) = \psi_2(b_2) = \psi_2\beta(a_2) = \gamma\phi_2(a_2).$$

Hence, $b_3 = \gamma(a_3 + \phi_2(a_2))$. Thus, γ is an epimorphism.

It should be noted that the steps in the proof and their arrangement within (1) and (2) cannot be avoided. Of the hypotheses, the argument only required that β and δ be isomorphisms, α an epimorphism, and ϵ a monomorphism.

5.28. Application: *Let $f:(X, A) \longrightarrow (Y, B)$ be continuous. Let $f_1:X \longrightarrow Y$ and $f_2:A \longrightarrow B$ be the functions induced by f. If two of f_*, f_{1*}, f_{2*} are isomorphisms for all q, then the third is also.* This follows from application of the "Five" Lemma to the diagram

$$\begin{array}{ccccccccccccccc}
\to H_{q+1}(X) & \to & H_{q+1}(X,A) & \to & H_q(A) & \to & H_q(X) & \to & H_q(X,A) & \to & H_{q-1}(A) & \to & H_{q-1}(X) & \to \\
\downarrow f_{1*} & & \downarrow f_* & & \downarrow f_{2*} & & \downarrow f_{1*} & & \downarrow f_* & & \downarrow f_{2*} & & \downarrow f_{1*} & \\
\to H_{q+1}(Y) & \to & H_{q+1}(Y,B) & \to & H_q(B) & \to & H_q(Y) & \to & H_q(Y,B) & \to & H_{q-1}(B) & \to & H_{q-1}(Y) & \to .
\end{array}$$

5.29. Application: Let $(Y, B) \subset (X, A)$. Let Y be a deformation retract of X, and let B be a deformation retract of A. The injection $f:(Y, B) \longrightarrow (X, A)$ induces injections $f_1: Y \longrightarrow X$ and $f_2:B \longrightarrow A$ for which f_{1*} and f_{2*} are isomorphisms. Hence, f_* is an isomorphism.

6

The Excision Theorem

Let E and F be convex subsets of affine spaces and let $\sigma = (a_0, a_1, \cdots, a_q)$ be an affine q-simplex of E. The point $b_\sigma = \sum_{i=0}^{q}(1/q+1)a_i$ is called the *barycenter* (center of gravity) of $|\sigma|$. For $\rho \in A_0(E, F)$,

$$b_\sigma^\rho = \sum_{i=0}^{q}\left(\frac{1}{q+1}\right)\alpha_i^\rho = b_{\rho\sigma}.$$

For reasons that will be seen later, the considerations of this chapter will be restricted to the non-augmented case.

Definitions: In the case of affine spaces, the homomorphism which is to be called a *subdivision*,

$$\mathrm{Sd}_q : A(\Delta_q, E) \longrightarrow A(\Delta_q, E),$$

is defined by $\mathrm{Sd}_q = \mathrm{Id}$ for $q \leq 0$ and $\mathrm{Sd}_q(\sigma) = b_\sigma \mathrm{Sd}_{q-1}\partial_q(\sigma)$ for $q > 0$ and $\sigma \in A_0(\Delta_q, E)$. Thus for a 0-simplex a_0, the definition states that $\mathrm{Sd}_0(a_0) = a_0$. For a 1-simplex $\sigma = (a_0, a_1)$, the definition yields

$$\mathrm{Sd}_1(a_0, a_1) = b_\sigma((a_1) - (a_0)) = (b_\sigma, a_1) - (b_\sigma, a_0);$$

pictorially,

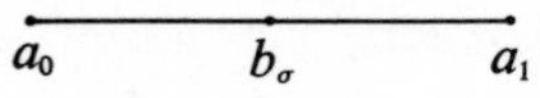

Figure 10

For a 2-simplex $\sigma = (a_0, a_1, a_2)$, $\mathrm{Sd}_2(\sigma)$ has the appearance

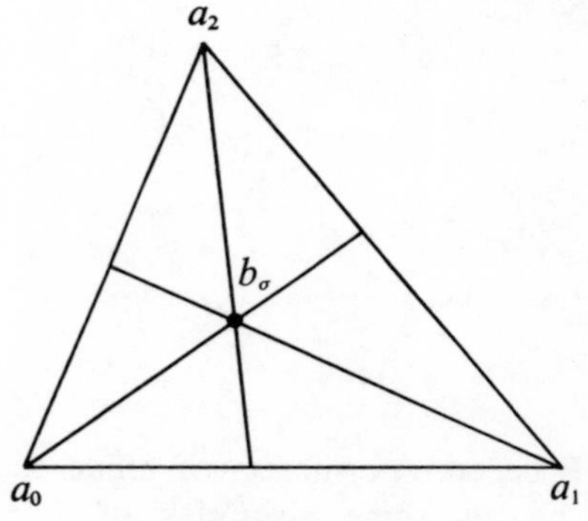

Figure 11

The homomorphism $R_q : A(\Delta_q, E) \longrightarrow A(\Delta_{q+1}, E)$ is defined recursively by $R_q = 0$ for $q \leq 0$, and

$$R_q(\sigma) = b_\sigma(\mathrm{Id} - \mathrm{Sd}_q - R_{q-1}\partial_q)(\sigma)$$

for $q > 0$ and $\sigma \in A_0(\Delta_q, E)$.

6.1. THEOREM: Sd *is a chain homomorphism; that is,*

(1) $$\mathrm{Sd}_{q-1}\partial_q = \partial_q \mathrm{Sd}_q,$$

and is chain homotopic to the identity because

(2) $$\partial_{q+1}R_q + R_{q-1}\partial_q = \mathrm{Id} - \mathrm{Sd}_q.$$

Proof: Because of $\partial_q = 0$, both formulas hold for $q \leq 0$ in the non-augmented case. Suppose that (1) has been proved for some $q \geq 0$. Then,

$$\partial_q \mathrm{Sd}_q \partial_{q+1} = \mathrm{Sd}_{q-1}\partial_q\partial_{q+1} = 0,$$

and hence,

$$\partial_{q+1}\mathrm{Sd}_{q+1}(\sigma) = \partial_{q+1}b_\sigma \mathrm{Sd}_q\partial_{q+1}(\sigma) = \mathrm{Sd}_q\partial_{q+1}(\sigma)$$

since 3.5 can be written $\partial_{q+1}b_\sigma = \mathrm{Id} - b_\sigma\partial_\sigma$. Now, suppose that (2) has been proved for a $q \geq 0$. Then,

$$\partial_{q+1}R_q\partial_{q+1} = (\mathrm{Id} - \mathrm{Sd}_q)\partial_{q+1} = \partial_{q+1}(\mathrm{Id} - \mathrm{Sd}_{q+1}),$$

and, since $\partial_{q+2}b_\sigma = \mathrm{Id} - b_\sigma\partial_{q+1}$, it follows that

$$\begin{aligned}\partial_{q+2}R_{q+1}(\sigma) &= \partial_{q+2}b_\sigma(\mathrm{Id} - \mathrm{Sd}_{q+1} - R_q\partial_{q+1})(\sigma)\\ &= (\mathrm{Id} - \mathrm{Sd}_{q+1} - R_q\partial_{q+1})(\sigma).\end{aligned}$$

6.2. *For $\rho \in A_0(E, F)$, both $\rho\mathrm{Sd}_q = \mathrm{Sd}_q\rho$ and $\rho R_q = R_q\rho$ hold.*

Proof: The assertion is correct for $q \leq 0$ because $\mathrm{Sd}_q = \mathrm{Id}$ and $R_q = 0$. If $\rho\mathrm{Sd}_q = \mathrm{Sd}_q\rho$ has already been proved for a $q \geq 0$, then it follows for $\sigma \in A_0(\Delta_{q+1}, E)$ that

$$\begin{aligned}\rho\mathrm{Sd}_{q+1}\sigma &= \rho b_\sigma\mathrm{Sd}_q\partial_{q+1}\sigma = b_\sigma^\rho\mathrm{Sd}_q\partial_{q+1}\rho\sigma\\ &= b_{\rho\sigma}\mathrm{Sd}_q\partial_{q+1}\rho\sigma = \mathrm{Sd}_{q+1}\rho\sigma.\end{aligned}$$

Thus, $\rho\mathrm{Sd}_{q+1} = \mathrm{Sd}_{q+1}\rho$. If $\rho R_q = R_q\rho$ has already been proved for $q \geq 0$, then

$$\begin{aligned}\rho R_{q+1}\sigma &= \rho b_\sigma(\mathrm{Id} - \mathrm{Sd}_{q+1} - R_q\partial_{q+1})\sigma\\ &= b_{\rho\sigma}(\mathrm{Id} - \mathrm{Sd}_{q+1} - R_q\partial_{q+1})\rho\sigma = R_{q+1}\rho\sigma.\end{aligned}$$

Thus, $\rho R_{q+1} = R_{q+1}\rho$.

E will now be specialized to Δ_q and σ to the identity, denoted by $\Delta_q^\dagger : \Delta_q \longrightarrow \Delta_q$ for $q \geq 0$. Then $\rho \in A_0(\Delta_q, E)$ is a q-simplex of F. Since $\rho\Delta_q^\dagger = \rho$, the formulas (1) and (2) of 6.1 yield

$$\rho\mathrm{Sd}_q(\Delta_q^\dagger) = \mathrm{Sd}_q(\rho) \text{ and } \rho R_q(\Delta_q^\dagger) = R_q(\rho).$$

For arbitrary $c \in A(\Delta_q, F)$ it then follows that

6.3. $$c\mathrm{Sd}_q(\Delta_q^\dagger) = \mathrm{Sd}_q(c) \text{ and } cR_q(\Delta_q^\dagger) = R_q(c).$$

For $c = \partial_q\Delta_q^\dagger$ and $q \geq 1$, it follows that

$$\partial_q(\Delta_q^\dagger)R_{q-1}(\Delta_{q-1}^\dagger) = R_{q-1}\partial_q(\Delta_q^\dagger).$$

According to 3.9,

$$R_q(\Delta_q^\dagger)\partial_{q+1}(\Delta_q^\dagger) = \partial_{q+1}R_q(\Delta_q^\dagger).$$

Addition and (2) of 6.1 yield

6.4. $$\partial_q(\Delta_q^\dagger)R_{q-1}(\Delta_{q-1}^\dagger) + R_q(\Delta_q^\dagger)\partial_{q+1}(\Delta_{q+1}^\dagger) = \Delta_q^\dagger - \mathrm{Sd}_q(\Delta_q^\dagger).$$

Definition: Let (X, A) be a pair of topological spaces. In the non-augmented case, the functions $\mathrm{Sd}_q: S_q(X, A) \longrightarrow S_q(X, A)$ and $R_q: S_q(X, A) \longrightarrow S_{q+1}(X, A)$ are defined by $\mathrm{Sd}_q = \mathrm{Id}$ and $R_q = 0$, for $q < 0$;

$$\mathrm{Sd}_q(c + S_q(A)) = c\mathrm{Sd}_q(\Delta_q^\dagger) + S_q(A),$$

for $q \geq 0$; and

$$R_q(c + S_q(A)) = cR_q(\Delta_q^\dagger) + S_{q+1}(A),$$

for $q \geq 0$. From

$$\mathrm{Sd}_0(\Delta_0^\dagger) = \mathrm{Id}(\Delta_0^\dagger) = \mathrm{Id} = \Delta_0^\dagger \text{ and } R_0(\Delta_0^\dagger) = 0$$

it follows that $\mathrm{Sd}_0 = \mathrm{Id}$ and $R_0 = 0$. By Lemma 1.1, Sd_q and R_q are well-defined homomorphisms.

Formula (2) of 6.1 will now be established in the general case. It is trivial for $q \leq 0$ since, then, $\partial_0 = 0$, $R_0 = 0$ and $\mathrm{Sd}_0 = \mathrm{Id}$. For $q \geq 1$, formula 6.5 yields

$$\begin{aligned}(\partial_{q+1}R_q + R_{q-1}\partial_q)(c + S_q(A)) &\\ &= c(R_q(\Delta_q^\dagger)\partial_{q+1}(\Delta_{q+1}^\dagger) + \partial_q(\Delta_q^\dagger)R_{q-1}(\Delta_{q-1}^\dagger) + S_q(A)\\ &= c(\Delta_q^\dagger - \mathrm{Sd}_q(\Delta_q^\dagger)) + S_q(A)\\ &= (\mathrm{Id} - \mathrm{Sd}_q)(c + S_q(A)).\end{aligned}$$

Hence, for topological space pairs,

6.1(2′). $$\partial_{q+1}R_q + R_{q-1}\partial_q = \mathrm{Id} - \mathrm{Sd}_q.$$

6.5. Lemma: *Let A and B be chain complexes. For each q, let a homomorphism $D_q:A_q \longrightarrow B_{q+1}$ be given. If f_q is defined by*

6.6. $$f_q = \partial_{q+1}D_q + D_{q-1}\partial_q,$$

then $f_q:A_q \longrightarrow B_q$ is a chain homomorphism for each q.

Proof: Clearly, f_q is well defined. From 6.6, it follows that

$$\partial_q f_q = \partial_q D_{q-1}\partial_q = f_{q-1}\partial_q.$$

6.7. Corollary: *If c_q is a cycle, then c_q is homologous to its subdivision* $\mathrm{Sd}_q c_q$.

Proof: Application of Lemma 6.5 to $f_q = \mathrm{Id} - \mathrm{Sd}_q = \partial_{q+1}R_q + R_{q-1}\partial_q$ shows that $\partial_q \mathrm{Sd}_q = \mathrm{Sd}_{q-1}\partial_q$. Hence, $\mathrm{Sd}:S(X, A) \longrightarrow S(X, A)$ is a chain homomorphism that is chain homotopic to the identity by 6.1(2′). By Theorem 5.4, $(\mathrm{Sd}_q)_* = \mathrm{Id}$.

6.8. Corollary: *Let $q \geq 0$ and let $\sigma \in C_0(\Delta_q, X)$ be a singular q-simplex. Then the carriers of $\partial_q\sigma$, $\mathrm{Sd}_q\sigma$ and $R_q\sigma$ lie in $|\sigma|$.*

Proof: For $a \in C(Y, \Delta_q)$, σa is in $C(Y, X)$. It is immediate that the carrier $|\sigma a|$ of σa (a pointset in X) is a subset of $|\sigma|$. Now choose $a = \partial_q(\Delta_q^\dagger)$, $a = \mathrm{Sd}_q(\Delta_q^\dagger)$ or $a = R_q(\Delta_q^\dagger)$, and the corollary follows.

At this point, there is a digression into properties of certain metric spaces.

Definition: By the *diameter* $\delta(M)$ of a non-empty set M in a space with metric $d(x, y)$ is meant

$$\delta(M) = \sup_{x, y \in M} d(x, y).$$

The metric of a euclidean vector space E is written $|x - y|$.

Let $\sigma = (a_0, a_1, \cdots, a_q) \in A_0(\Delta_q, E)$. The *carrier* $|\sigma|$ is the geometrical simplex with the vertices $a_0, a_1, \cdots, a_q$. Let $x, y \in |\sigma|$; that is, $x = \sum \xi_i a_i$, $y = \sum \eta_i a_i$, where $\xi_i \geq 0$, $\eta_i \geq 0$, $\sum \xi_i = \sum \eta_i = 1$, and $i = 0, 1, \cdots, q$. Then

$$|y - x| = |\sum \xi_i(y - a_i)| \leq \mathrm{Max}\,|y - a_i|,$$

and hence,

(a) $$|y - x| \leq \underset{i}{\text{Max}}\, |y - a_i| \leq \underset{i,j}{\text{Max}}\, |a_j - a_i|.$$

Since $a_i \in |\sigma|$, it follows that

(b) $$\delta(|\sigma|) = \underset{i,j}{\text{Max}}\, |a_i - a_j|.$$

More precisely,

$$|y - a_i| = |(\sum_j \eta_j)(a_j - a_i)| \leq (1 - \eta_i)\ \text{Max}\, |a_j - a_i|$$
$$= (1 - \eta_i)\delta(|\sigma|).$$

Then for

$$y = b_\sigma = \sum_i (1/(q + 1))a_i$$

it follows that $|b_\sigma - a_i| \leq (q/q + 1)\delta(|\sigma|)$, and hence, according to (a),

(c) $$|b_\sigma - x| \leq \frac{1}{q + 1}\,\delta(|\sigma|).$$

6.9. THEOREM: *For each q-simplex τ of* $\text{Sd}_q\sigma$,

$$\delta(\tau) \leq \frac{q}{q + 1}\,\delta(|\sigma|).$$

Proof (by induction): For $q = 0$, the conclusion is correct since $\delta(|\tau|) = 0 = \delta(|\sigma|)$. Suppose that it has been proved for $q - 1$. Since $\text{Sd}_q\sigma = b_\sigma \text{Sd}_{q-1}\partial_q$, there is to τ a simplex α of $\partial_q\sigma$ and a simplex β of $\text{Sd}_{q-1}\alpha$ with $\tau = b_\sigma\beta$. Since $|\alpha| \subset |\sigma|$, it follows that $\delta(|\alpha|) \leq \delta(|\sigma|)$. Therefore, by inductive hypothesis,

$$\delta(|\beta|) \leq ((q - 1)/q)\delta(|\sigma|) \leq ((q + 1)/q) \sum (|\sigma|).$$

From (c), it follows that

$$|b_\sigma - x| \leq (q/(q + 1))\delta(|\sigma|)$$

when $x \in |\beta| \subset |\sigma|$. Hence, (b) implies that

$$\delta(|\tau|) \leq (q/(q + 1))\cdot(|\sigma|).$$

Remark: The estimation of $\delta(|\tau|)$ may appear to be coarse, but it is not. For let $a_0 = a_1 = \cdots = a_{q-1} = 0$, $a_q \neq 0$. Then $b_\sigma = (q/(q+1))a_q$ and

$$|a_q - b_\sigma| = (q/(q+1))|a_q| = (q/(q+1))\delta(|\sigma|).$$

A simplex τ that has a_q as vertex will then satisfy

$$\delta(|\tau|) = (q/(q+1))\delta(|\sigma|).$$

Suppose that the subdivision Sd has been carried out repeatedly. For each affine simplex τ that appears in $\mathrm{Sd}_q^n(\sigma)$ with non-zero coefficient, it follows that

$$\delta(|\tau|) \leq (q/(q+1))^n\delta(|\sigma|).$$

Hence, there is to each real $\epsilon > 0$ an n such that all simplices of $\mathrm{Sd}_q^n(\sigma)$ have a diameter $< \epsilon$.

Now return to formula 6.3: For $c \in A(\Delta_q, F)$,

$$\mathrm{Sd}(c) = c\,\mathrm{Sd}(\Delta_q^\dagger) \in A(\Delta_q, F),$$

where $q \geq 0$ and Sd is written in place of Sd_q. Iteration yields $\mathrm{Sd}^n c = c(\mathrm{Sd}(\Delta_q^\dagger))^n$, for $n = 1, 2, \cdots$ where the symbol on the left of the equality stands for the nth subdivision of c and the symbol on the right stands for the product of n functions. This formula also holds for $n = 0$ when $\mathrm{Sd}^0 c$ is defined to be c. For $F = \Delta_q$ and $c = \Delta_q^\dagger = \mathrm{Id}$, it follows that

6.10. $$\mathrm{Sd}^n(\Delta_q^\dagger) = (\mathrm{Sd}(\Delta_q^\dagger))^n.$$

Definition: Let (X, A) be a pair of topological spaces. For $q \geq 0$, let

$$\mathrm{Sd}(c + S_q(A)) = c\mathrm{Sd}(\Delta_q^\dagger) + S_q(A).$$

Then, by 6.10,

$$\mathrm{Sd}^n(c + S_q(A)) = c(\mathrm{Sd}\Delta_q^\dagger)^n + S_q(A) = c\mathrm{Sd}^n(\Delta_q^\dagger) + S_q(A).$$

The standard simplex Δ_q is a compact, metric space because, for instance, it is closed and bounded. The following theorem will have application to simplices:

6.11. LEBESGUE COVERING THEOREM: *Let M be a compact, metric space. To each covering of M by open sets B, there is an $\epsilon > 0$ such that each subset of M whose diameter is less than ϵ lies entirely in one of the sets B.*

Proof: Each $x \in M$ lies in a B. Since B is open, there is an open ball around x of radius r_x which is contained in B. Let K_x be the open ball around x of radius $\frac{1}{2}r_x > 0$. M is covered by the set of all K_x. Since M is compact, M is covered by a finite collection $\{K_{x_i}\}$, $i = 1, 2, \cdots, n$. Let $\epsilon = \text{Min}\, r_{x_i}$, $A \subset M$, and $\delta(A) < \epsilon$. If $x \in A$, then $x \in K_{x_i}$, for some i, and $d(x, x_i) < \frac{1}{2}r_{x_i}$. For $y \in A$, $d(x, y) < \epsilon \leq \frac{1}{2}r_{x_i}$. Hence, $d(y, x_i) < \frac{1}{2}r_{x_i} + \frac{1}{2}r_{x_i} = r_{x_i}$. Therefore, all of A lies in the ball of radius r_{x_i} around x_i and hence in B.

Now consider $c\text{Sd}^n(\Delta_q^\dagger)$ for the case of a singular simplex $c = \sigma \in C_0(\Delta_q, X)$. Each affine simplex τ which appears in $\text{Sd}^n(\Delta_q^\dagger)$ with a coefficient different from zero is a mapping of Δ_q upon $|\tau| \subset \Delta_q$. Hence, the carrier of $\sigma\tau:\Delta_q \longrightarrow X$ is the image of $|\tau|$ under σ.

Let $\mathring{B}$ denote the interior of the set B. Let $\mathscr{B}$ be a covering of X with sets B so that the collection of open sets $\mathring{B}$ covers X. Since $\sigma:\Delta_q \longrightarrow X$ is a continuous function (here for the first time the continuity of the map σ is used), each of the sets $\sigma^{-1}(\mathring{B})$ is open; and the collection of these sets is a covering of Δ_q. Since Δ_q is a compact, metric space, there is an $\epsilon > 0$ (called the lebesgue number) such that each subset of Δ_q with diameter less than ϵ is contained in one of the sets $\sigma^{-1}(\mathring{B})$. For this ϵ choose an n such that $\delta|\tau| < \epsilon$ for each τ that occurs in $\text{Sd}^n(\Delta_q)$ with coefficient different from zero. Then each $|\tau|$ is in $\sigma^{-1}(\mathring{B})$ for some B, and hence, $|\sigma\tau| \subset \mathring{B} \subset B$.

When $c \in C(\Delta_q, X)$, the symbol $c \in \mathscr{B}$ will be used to indicate that to each simplex σ occurring in c there is a $B \in \mathscr{B}$ such that $|\sigma| \subset B$. It has just been proved that *to each singular simplex σ there is an n with* $\text{Sd}^n(\sigma) \in \mathscr{B}$. Here, as before, $\text{Sd}^0(\sigma) = \sigma$.

With each singular q-simplex σ of X there will be associated a natural number $n(\sigma)$ which satisfies the following conditions:

(a) For $\sigma \in \mathscr{B}$, $n(\sigma) = 0$.

(b) For each simplex σ_i which appears in $\partial_q(\sigma)$ with non-vanishing coefficient, $n(\sigma_i) \leq n(\sigma)$.

(c) $\text{Sd}^{n(\sigma)}(\sigma) \in \mathscr{B}$.

It can be seen by induction that such an association is possible. Indeed let $n(\sigma) = 0$ for all $(q - 1)$-simplices. For a q-simplex $\sigma \in \mathscr{B}$, all $\sigma_i \in \mathscr{B}$ and are $(q - 1)$-dimensional. By (a), $n(\sigma_i) = 0$ for all these σ_i; and hence, $n(\sigma) = 0$ satisfies (a), (b), and (c). To each q-simplex σ there is an n with $\mathrm{Sd}^n(\sigma) \in \mathscr{B}$. If $\sigma \notin \mathscr{B}$, choose $n(\sigma)$ larger than any of the finitely many $n(\sigma_i)$, where $\sigma_i \in \partial_q \sigma$. In the following, a fixed choice of $n(\sigma)$ is made for each σ.

In the non-augmented case, the homomorphism $D_q : S_q(X) \longrightarrow S_{q+1}(X)$ is defined for $q \geq 0$ by

6.12. $$D_q(\sigma) = \sum_{0 \leq k \leq n(\sigma)} R_q \mathrm{Sd}_q^k(\sigma).$$

For $q < 0$, D_q is defined by $D_q = 0$. For $n(\sigma) = 0$, in particular for $q = 0$, the empty sum is to have the value 0; and then $D_q = 0$.

For $q \geq 1$ and $\sigma \in S_q(X)$, Formula 6.1(2′) and $\partial_q \mathrm{Sd}_q = \mathrm{Sd}_{q-1} \partial_q$ yield

$$\begin{aligned}\partial_{q+1} D_q(\sigma) &= \sum_{0 \leq k < n(\sigma)}^{q} (\mathrm{Id} - \mathrm{Sd}_q - R_{q-1}\partial_q)\mathrm{Sd}_q^k(\sigma) \\ &= \sigma - \mathrm{Sd}_q^{n(\sigma)}(\sigma) - \sum_{0 \leq k < n(\sigma)} R_{q-1}\mathrm{Sd}_{q-1}^k \partial_q(\sigma).\end{aligned}$$

From $\partial_q(\sigma) = \sum_{a_i \neq 0} a_i \sigma_i$, it follows that

$$\partial_{q+1} D_q(\sigma) = \sigma - \mathrm{Sd}_q^{n(\sigma)}(\sigma) - \sum a_i \sum_{0 \leq k < n(\sigma)} R_{q-1}\mathrm{Sd}_{q-1}^k \sigma_i.$$

On the other hand, 6.12 yields

$$D_{q-1}\partial_q(\sigma) = \sum a_i D_{q-1}(\sigma_i) = \sum a_i \sum_{0 \leq k < n(\sigma_i)} R_{q-1}\mathrm{Sd}_{q-1}^k(\sigma_i).$$

Because $n(\sigma) \geq n(\sigma_i)$, addition yields

$$\begin{aligned}&\partial_{q+1} D_q(\sigma) + D_{q-1}\partial_q(\sigma) \\ &\qquad = \sigma - \mathrm{Sd}_q^{n(\sigma)}(\sigma) - \sum_{a_i \neq 0} a_i \Big(\sum_{n(\sigma_i) \leq k < n(\sigma)} R_{q-1}\mathrm{Sd}_{q-1}^k(\sigma_i)\Big).\end{aligned}$$

The homomorphism $\tau_q : C(\Delta_q, X) \longrightarrow C(\Delta_q, X)$ is now defined by

6.13. $$\tau_q(\sigma) = \mathrm{Sd}^{n(\sigma)}(\sigma) + \sum_{a_i \neq 0} a_i \Big(\sum_{n(\sigma_i) \leq k < n(\sigma)} R_{q-1}\mathrm{Sd}_{q-1}^k(\sigma_i)\Big).$$

Then,

$$\partial_{q+1} D_q + D_{q-1}\partial_q = \mathrm{Id} - \tau_q \text{ for } q \geq 1.$$

For $q \leq 0$, $D_q = 0$; this means that defining $\tau_q = \mathrm{Id}$, when $q \leq 0$, defines $\tau_q : S_q(X) \longrightarrow S_q(X)$ for all q in such a way that

6.14. $$\partial_{q-1} D_q + D_{q-1}\partial_q = \mathrm{Id} - \tau_q.$$

6.15. Properties of τ_q:

(1) From Lemma 6.5, it follows that

$$\partial_q \tau_q = \tau_{q-1} \partial_q.$$

(2) For $q \geq 0$ and each σ, $\tau_q(\sigma) \in \mathscr{B}$. This is trivial for $q = 0$. For $q > 0$, it follows from the fact that $\mathrm{Sd}_{q-1}^k(\sigma_i) \in \mathscr{B}$ for $k \geq n(\sigma_i)$.

(3) For $\sigma \in \mathscr{B}$, $\tau_q(\sigma) = \mathrm{Sd}_q^0(\sigma) = \sigma$.

Notation: For $q \geq 0$, the module of all q-chains $c \in S_q(X)$ for which $c \in \mathscr{B}$ will be designated by $T_q(X)$. For $q < 0$, T_q is defined to be 0.

Let $T_q(X)$ be a submodule of $S_q(X)$. For $A \subset X$, $T_q(A) \subset S_q(A)$. Hence, if $c \in \mathscr{B}$, then $\partial_q c \in \mathscr{B}$. Therefore, $\partial_q T_q(X) \subset T_{q-1}(X)$ and $\partial_q T_q(A) \subset T_{q-1}(A)$.

The factor group $T_q(X)/T_q(A)$ will be designated by $T_q(X, A)$. The operator ∂_q induces a boundary operator $\partial_q : T_q(X, A) \longrightarrow T_{q-1}(X, A)$.

From 6.15(2), it follows that

6.16. $\tau_q(S_q(X)) \subset T_q(X)$ and $\tau_q(S_q(A)) \subset T_q(A)$.

Definition: The homomorphism $\lambda_q : S_q(X, A) \longrightarrow T_q(X, A)$ is defined by

$$\lambda_q(c + S_q(A)) = \tau_q c + T_q(A).$$

Lemma 1.1 and Theorem 6.1(2) guarantee that λ_q is well defined.

Definition: The injection-type homomorphism $j_q : T_q(X, A) \longrightarrow S_q(X, A)$ is defined by

$$j_q(c + T_q(A)) = c + S_q(A) \text{ for } c \in T_q(X).$$

From 6.15(1), it follows that $\lambda_{q-1}\partial_q = \partial_q \lambda_q$. For j_q, it can be checked directly that $j_{q-1}\partial_q = \partial_q j_q$. Hence, both λ and j are chain homomorphisms. By 6.15(3), it is clear that $\lambda_q j_q = \mathrm{Id}$ and $j_q \lambda_q = \tau_q$.

Since $\lambda_j = \mathrm{Id}$ and $\mathrm{Id} - j\lambda = \partial D + D\partial$, Theorem 5.4 can be applied to the chain homomorphisms j and λ. Hence,

$$(j\lambda)_* = j_* \lambda_* = \mathrm{Id}$$

and, also,

$$\lambda_* j_* = (\lambda j)_* = \mathrm{Id}.$$

Therefore, λ_* *and* j_* *are isomorphisms for the corresponding homology groups.*

In a geometrical interpretation of the foreogoing, it should be noted that a covering $\mathscr{B}$ of X leads to a notion of a "small simplex"; namely, a simplex whose carrier is contained in some element of $\mathscr{B}$. In this sense, $T(X, A)$ is the chain complex which has been generated by the small simplices only, and the map $j:T(X, A) \longrightarrow S(X, A)$ induces an isomorphism of the homology groups of $T(X, A)$ onto the homology groups of $S(X, A)$. This can be applied in proving the next theorem.

6.17. THE EXCISION THEOREM: *Let (X, A) be a space pair. Let U be a set in X such that $\bar{U} \subset \mathring{A}$. Let i be the injection $i:(X - U, A - U) \longrightarrow (X, A)$. Then, in the non-augmented homology, $i_*: H_q(X - U, A - U) \longrightarrow H_q(X, A)$ is an isomorphism.*

Proof: The covering $\mathscr{B}$ used here will simply consist of the two sets: A and $X - U$. The interior of $X - U$ is $X - \bar{U}$; and, since $\bar{U} \subset \mathring{A}$, X is also covered by the two sets $\mathring{A}$ and $\mathrm{Int}(X - U)$. For this $\mathscr{B}$, the group $T_q(X)$ will now be computed: A simplex σ lies in $T_q(X)$ if $|\sigma|$ is contained either in A or in $X - U$. Hence,

$$T_q(X) = S_q(A) + S_q(X - U),$$

a statement that is trivial for $q < 0$. This sum is not necessarily direct. It is trivial that $T_q(A) = S_q(A)$, and therefore the isomorphism

$$T_q(X, A) \cong \frac{S_q(X - U) + S_q(A)}{S_q(A)}$$

follows.

For the next part of the proof, the Module Isomorphism Theorem is needed; this states that *if* M_1 and M_2 *are submodules of a larger module, and* if

$$k: \frac{M_2}{M_1 \cap M_2} \longrightarrow \frac{(M_1 + M_2)}{M_1}$$

is defined by $k(m_2 + M_1 \cap M_2) = m_2 + M_1$, *then* k *is an isomorphism*. In the application here, M_1 is $S_q(X - U)$ and M_2 is $S_q(A)$. Since these modules are free, the intersection $S_q(X - U) \cap S_q(A)$ is generated by the intersection of the respective sets of generators:

for $q \geq 0$, by those simplices σ whose carrier $|\sigma|$ lies both in $X - U$ and in A. This is the same as saying that $|\sigma| \subset A - U$. Hence,

$$S_q(X - U) \cap S_q(A) = S_q(A - U).$$

This is trivial for $q < 0$, since in that case the modules are 0. Now let $k: S_q(X - U, A - U) \longrightarrow T_q(X, A)$ be defined by

$$k(c + S_q(A - U)) = c + S_q(A) \qquad \text{for } c \in S_q(X - U).$$

Then, by the Module Isomorphism Theorem, k is an isomorphism. Obviously $k\partial = \partial k$, and therefore k is a chain homomorphism. The following are also chain homomorphisms:

$$S(X, A) \xrightarrow{\lambda} T(X, A),$$
$$T(X, A) \xrightarrow{k^{-1}} S(X - U, A - U),$$

and

$$S(X - U, A - U) \xrightarrow{k} T(X, A) \xrightarrow{j} S(X, A).$$

Hence,

$$jk(c + S_q(A - U)) = j(c + S_q(A)) = c + S_q(A).$$

On the other hand, the injection $i:(X - U, A - U) \longrightarrow (X, A)$ induces a chain homomorphism $i^{\#}: S(X - U, A - U) \longrightarrow S(X, A)$ in accordance with Theorem 4.6. Since

$$i^{\#}(c + S_q(A - U)) = ic + S_q(A) = c + S_q(A),$$

it follows that $jk = i^{\#}$. Let $\rho = k^{-1}\lambda$. Then

$$\rho: S(X, A) \longrightarrow S(X - U, A - U),$$
$$\rho i^{\#} = k^{-1}\lambda jk = k^{-1}\mathrm{Id}\, k = \mathrm{Id},$$

and

$$i^{\#}\rho = jkk^{-1}\lambda = j\lambda.$$

Consequently,

$$\mathrm{Id} - j\lambda = \partial D + D\partial.$$

Theorem 5.4 yields

$$(i^{\#}\rho)_* = i_*\rho_* = \text{Id}.$$

From $\rho i^{\#} = \text{Id}$, it follows that $\rho_* i_* = \text{Id}$; therefore, i_* is an isomorphism.

Notice that U did not need to be an open set here.

Remark on the augmented case: By definition, $H_q(X, A) = \text{Ker } \partial_q / \text{Im } \partial_{q+1}$ where $\partial_q : S_q(X, A) \longrightarrow S_{q-1}(X, A)$. For $q > 0$, the augmented groups $S_q(X, A)$ and the homomorphisms ∂_q are identical with those that occur in the non-augmented case. Hence, the groups $H_q(X, A)$ are the same in both cases, provided that $q > 0$ or $q = 0$ and $A \neq \varnothing$. For $q < 0$, $H_q(X, A) = 0$ by Theorem 4.4. There is then a difference between the augmented and the non-augmented homology groups only in the case $X \neq \varnothing$, $A = \varnothing$, $q = 0$; that is, in the computation of $H_0(X, \varnothing) = H_0(X)$. However, the Excision Theorem is meaningful only when $A \neq \varnothing$. Therefore, the Excision Theorem can also be employed in the augmented case except when $q = 0$ and $A - U = \varnothing$. This exceptional case seldom arises in applications and, because of $\overline{U} \subset \mathring{A}$, can only occur when A is both open and closed.

7

Direct Decomposition and Additional Aids to the Computation of Homology Groups

7.1. *Let G be the direct sum of R-modules G_k, $k \in J$. For each k, let H_k be a submodule of G_k. Then the sum H of the H_k is direct. The direct sum $\sum^{\oplus} G_k/H_k$ is isomorphic to G/H.*
Proof:
Let

$$i(\textstyle\sum(g_k + H_k)) = \sum g_k + H_k \text{ and } j(\sum g_k + H) = \sum(g_k + H_k)$$

define homomorphisms

$$i{:}\textstyle\sum^{\oplus} G_k/H_k \longrightarrow G/H \text{ and } j{:}G/H \longrightarrow \sum^{\oplus} G/_k H_k.$$

Since the sums are direct, i and j are well defined. Since $ij = \text{Id}$ and

$ji = \text{Id}$, i is an isomorphism. Let $i_k:G_k/H_k \longrightarrow G/H$ be defined by $i_k(g_k + H_k) = g_k + H$. The situation will be indicated by writing $i = \sum i_k$ *directly*.

Now let (X, A) be a topological space pair. Let $X = \cup X_k$ be a disjoint decomposition of X into subsets X_k, and for each k and each $x \in X_k$ let X_k contain the *path-component* of x; that is, the set of points connectible to x by a path. Here it should be recollected that a path is a singular 1-simplex. For $A_k = A \cap X_k$, the union $A = \cup A_k$ is disjoint. The injections $i_k(X_k, A_k) \longrightarrow (X, A)$ induce homomorphisms $\sum(i_k)_*: \sum^{\oplus} H_q(X_k, A_k) \longrightarrow H_q(X, A)$.

7.2. THEOREM: *In the non-augmented case, $\sum(i_k)_*$ is an isomorphism.*

Proof: Let $\sigma:\Delta_q \longrightarrow X$. The set $|\sigma|$ is the continuous image of Δ_q and is therefore pathwise connected. Hence it is contained in precisely one X_k. Consequently,

$$S_q(X) = \sum_k{}^{\oplus} S_q(X_k) \text{ and } S_q(A) = \sum_k{}^{\oplus} S_q(A), \text{ for } q \geq 0.$$

By definition, $S_q(X, A) = S_q(X)/S_q(A)$ and $S_q(X_k, A_k) = S_q(X_k)/S_q(A_k)$. For $i_k^{\#}:S_q(X_k, A_k) \longrightarrow S_q(X, A)$, 7.1 shows that $\sum i_k^{\#}: \sum^{\oplus} S_q(X_k, A_k) \longrightarrow S_q(X, A)$ is an isomorphism. The sum $\sum^{\oplus} S_q(X_k, A_k)$ forms a chain complex with ∂ defined for each component. Since the $i_k^{\#}$ commute with ∂, the sum $\sum i_k^{\#}$ does also. Direct computation of the image of an element of $H_q(\sum^{\oplus} S_q(X_k, A_k))$ under

$$(\sum i_k^{\#})_*:H_q(\sum^{\oplus} S_q(X_k, A_k)) \longrightarrow H_q(X, A)$$

in analogy to the proof of 7.1 shows that $(\sum i_k^{\#})_*$ is an isomorphism. By 7.1, there is also an isomorphism of $\sum^{\oplus} H_q(X_k, A_k)$ on $H_q(\sum^{\oplus} S_q(X_k, A_k))$. It is easy to check that the composition is an isomorphism

$$\sum(i_k^{\#})_*: \sum^{\oplus} H_q(X_k, A_k) \longrightarrow H_q(X, A).$$

For $q < 0$, the theorem is trivial in the non-augmented case. In the augmented case, the theorem is false since $S_{-1}(X, A)$ can be different from $\sum^{\oplus} S_{-1}(X_k, A_k)$; for instance, when $A = \varnothing$ and X has several path components.

If $\{X_k\}$ is a decomposition of X into its path components, then Theorem 7.2 permits the computation of $H_q(X, A)$ to be reduced to computation of the groups $H_q(X_k, A_k)$.

Computation of $H_0(X,A)$

Let r be the cardinality of the set of path components X_k of X for which $X_k \cap A = \varnothing$.

7.3. THEOREM: *$H_0(X, A)$ is a free r-module. It has r basis elements when $A \neq \varnothing$ or when the group is non-augmented and $A = \varnothing$. It has r — 1 basis elements when the group is augmented and $A \neq \varnothing$.* This theorem yields the following corollary.

7.4. Corollary: *By choice of a suitable ring R, for instance $R = Z$ or $R = Q$, the number of path-components of X can be determined from $H_0(X)$. In any case, for an arbitrary ring R, X is pathwise connected only if in the augmented case $H_0(X) = 0$.*

Proof of 7.3:

(1) The non-augmented case. By 7.2, X may be taken to be pathwise connected. From $\partial_0 = 0$, it follows that each element of $S_0(X, A)$ is a cycle. The bounding cycles will now be determined:

(a) Let $A \neq \varnothing$ and let $P \in A$. Each $Q \in X$ can be connected to P by a path σ. Then

$$\partial\sigma = \sigma\partial(\Delta_1^\dagger) = \sigma(d_1) - \sigma(d_0) = (Q) - (P),$$

where (Q) denotes the 0-simplex $d_0 \longrightarrow Q$. Furthermore,

$$\partial(\sigma + S_1(A)) = \partial\sigma + S_0(A) = (Q) - (P) + S_0(A) = (Q) + S_0(A).$$

The cosets $(Q) + S_0(A)$ generate $S_0(X, A)$. Therefore, each element of $S_0(X, A)$ is a boundary, and $H_0(X, A) = 0$.

(b) Let $A = \varnothing$. Let $c = \sum a_i\sigma_i$ be a 1-chain. Denote the beginning and end points of σ_i by P_i and Q_i, respectively. Then $\partial_1 c = \sum a_i((Q_i) - (P_i))$. A 0-chain $d = \sum b_i(R_i) \in S_0(X)$ is therefore a boundary only if $\sum b_i = 0$. Conversely, let $d = \sum b_i(R_i) \in S_0(X)$ and let $\sum b_i = 0$. Choose $P \in X$. Then

$$d = \sum b_i((R_i) - (P)) = \partial(\sum b_i\sigma_i),$$

where σ_i is a path from R_i to P. The association $d \longrightarrow \sum b_i$ defines a homomorphism of $S_0(X)$ on R whose kernel consists precisely of the boundaries. Hence, $H_0(X) \cong R$.

(2) The augmented case.

(a) Let $A \neq \varnothing$. Then by 6.18, the $H_q(X, A)$ are the same in both the augmented and non-augmented cases. They therefore have the same number of basis elements.

(b) Let $A = \varnothing$. The proof can be carried out either by a direct computation of cycles and boundaries or else as follows: The group $S_0(P)$ of a point P in the augmented case is generated by (P). From $\partial a(P) = a(.)$, $a \in R$, it follows that only the zero element is a cycle. Therefore, $H_0(P) = 0$. Now let $X \neq \varnothing$ be an arbitrary space and let $P \in X$. The exact sequence

$$0 = H_0(P) \longrightarrow H_0(X) \longrightarrow H_0(X, P) \longrightarrow H_{-1}(P) = 0$$

shows that $H_0(X)$ is isomorphic to $H_0(X, P)$. However, (X, P) was a case considered in (2a), and P lies in exactly one path-component of X. Therefore the number of basis elements is $r - 1$.

Homology Groups of a Point P

Theorem 7.3 shows that $H_0(P) = 0$ in the augmented case and that $H_0(P) \cong R$ in the non-augmented case. The next theorem handles all other dimensions.

7.5. THEOREM: *For $q \neq 0$, $H_q(P) = 0$.*

Proof: The result is already known for $q \leq 0$. For $q > 0$, the only singular simplex in P is $\sigma_q : \Delta_q \longrightarrow P$ and $S_q(P)$ is generated by it. The boundary $\partial_q \sigma_q$ is an alternating sum of $q + 1$ terms, each of which is a $(q - 1)$-simplex; that is, each term is σ_{q-1}. Consequently, $\partial_q \sigma_q = 0$ for q odd and $\partial_q \sigma_q = \sigma_{q-1}$ for q even. When q is odd, $\sigma_q = \partial \sigma_{q+1}$ and each element of $S_q(P)$ is a boundary. Hence, $H_q(P) = 0$. When q is even, $\partial_q(a\sigma_q) = a\sigma_{q-1}$ and $a\sigma_q$ is a cycle only if $a = 0$. Hence, $H_q(P) = 0$.

7.6. Corollary: *If X is homotopic to a point, then in the augmented case $H_q(X) = 0$ for all q.*

Examples:

(1) Let X be the set of rational numbers in the usual topology. Then the individual points form the path-components. Hence, $H_q(X) = 0$ for $q \neq 0$. $H_0(X)$ is a free module with countably many generators.

(2) Let X be the set of points $(x, y) \in R^2$ with $xy \neq 0$; in other words, the plane without the axes. Each of the four quadrants is homotopic to a point by means of a deformation retraction. Hence, $H_q(X) = 0$ for $q \neq 0$ and $H_0(X) \cong R + R + R$ in the augmented case.

Definition: Let (X, A) be a topological space pair and let $U \subset A$. Then the injection $i:(X - U, A - U) \longrightarrow (X, A)$ is called an *excision* if and only if, in the non-augmented case, $i_*:H_q(X - U, A - U) \longrightarrow H_q(X, A)$ is an isomorphism for each q. Notice that no assumption is made on the openness of U.

For instance, i is an excision when $\overline{U} \subset \mathring{A}$ (Excision Theorem).

7.7. Lemma: *Let $V \subset U \subset A$, and suppose that V defines an excision by means of the injection $(X - V, A - V) \longrightarrow (X, A)$.* Furthermore, let $(X - U, A - U)$ be a deformation retract of $(X - V, A - V)$. *Then U defines an excision.*

Proof: Consider the injections $i_1:(X - U, A - U) \longrightarrow (X - V, A - V)$ and $i_2:(X - V, A - V) \longrightarrow (X, A)$. By the comments preceding 5.26, $(i_1)_*$ is an isomorphism. Since i_2 is an excision, $(i_2)_*$ is an isomorphism. Hence, for $i = i_2 i_1$, the induced homomorphism $i_* = (i_2)_*(i_1)_*$ is an isomorphism.

An immediate consequence of the properties of exact sequences is the following lemma:

7.8. Lemma: *In the exact sequence of the triple (X, A, B)*

$$\cdots \longrightarrow H_q(A,B) \xrightarrow{i_*} H_q(X,B) \xrightarrow{j_*} H_q(X,A) \xrightarrow{\partial_*} H_{q-1}(A,B) \longrightarrow \cdots$$

let $H_q(\cdot,\cdot) = 0$ and $H_{q+1}(\cdot,\cdot) = 0$ for one of the space pairs. Then the two intermediate homology groups are isomorphic.

7.9. Corollaries:

(1) *If A is homotopic to a point and thus $H_q(A) = 0$ for all q in the augmented case, then $H_q(X)$ is isomorphic to $H_q(X, A)$ for all q.*

(2) *If X is homotopic to a point, then $\partial_*: H_q(X, A) \longrightarrow H_{q-1}(A)$ is an isomorphism in the augmented case.*

7.10. Lemma: *Let M and N be R-modules and let $j:M \longrightarrow N$ be an isomorphism. Let $g:M \longrightarrow M$ be defined by $g(x) = mx$ for a fixed m in R. Let the diagram*

$$\begin{array}{ccc} M & \xrightarrow{j} & N \\ {\scriptstyle g}\downarrow & & \downarrow{\scriptstyle h} \\ M & \xrightarrow[j]{} & N \end{array}$$

be commutative. Then $h(y) = my$ for $y \in N$.

Proof: Clearly, $h(y) = jgj^{-1}(y) = j(mj^{-1}(y)) = mjj^{-1}(y) = my$.

The Sphere

In euclidean space R^{n+1} the n-dimensional sphere S^n is defined by

$$S^n = \{x \mid x \in R^{n+1}, x_0^2 + \cdots + x_n^2 = 1\}.$$

The northern hemisphere E_+^n is defined by

$$E_+^n = \{x \mid x \in S_n^n, x_n \leq 0\},$$

the southern hemisphere E_-^n by

$$E_-^n = \{x \mid x \in S^n, x_n \leq 0\},$$

the equator by

$$E_+^n \cap E_-^n = \{x \mid x \in S^n, x_n = 0\},$$

and the equatorial n-cell E^n by

$$E^n = \{x \mid x \in R^{n+1}, |x| \leq 1, x_n = 0\}.$$

Clearly, $S^n = E_+^n \cup E_-^n$, and the equator can be identified with S^{n-1}. Furthermore, $S^{-1} = \varnothing$; $S^0 = \{+1, -1\}$, $E_+^0 = \{+1\}$; $E_-^0 = 0$, S^1 is a circle, and $E^1 = \{x \mid x_1 = 0, |x_0| \leq 1\}$. E^n is also referred to as an n-ball.

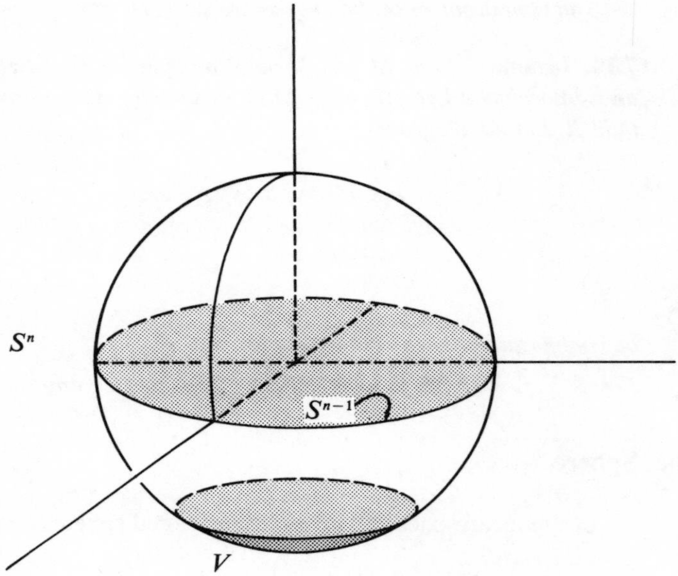

Figure 12

7.11. Lemma: *The injection* $\epsilon:(E^n, S^{n-1}) \longrightarrow (S^n, E^n_-)$ *is an excision.*

Proof: Remove from the n-sphere S^n a neighborhood, $V = \{x \mid x \in S^n, x_n < -\frac{1}{2}\}$, of the south pole. Since it is the intersection of S^n with a half-space, V is open. By the Excision Theorem, $(S^n - V, E^n_- - V) \longrightarrow (S^n, E^n_-)$ is an excision since $\overline{V} \in \mathring{E}^n_-$. Let $U = \mathring{E}^n_-$. It is sufficient to prove that $(S^n - U, E^n_- - U)$ is a deformation retract of $(S^n - V, E^n_- - V)$. For this purpose, define the projection function by $p(x_0, \cdots, x_n) = (x_0, \cdots, x_{n-1}, 0)$, and for $0 \leq t \leq 1$ and $x \in S^n - V$ define $F(x, t)$ by

$$F(x, t) = \begin{cases} x, \text{ for } x \in E^n_+ \\ \dfrac{(1-t)x + t\,p(x)}{|(1-t)x + t\,p(x)|}, \text{ for } x \in E^n_- - V. \end{cases}$$

For $x \in E^n_+ \cap (E^n_- - V) = S^n$, both $p(x) = x$ and $|x| = 1$. Hence, $F(x, t)$ is well defined. The denominator is not zero since the vectors

x and $p(x)$ have different directions for $x \notin S^{n-1}$. Hence, F is continuous. For $x \in E^n_- - V$, the numerator is a point of the line segment joining x with $p(x)$. Division by the absolute value induces a radial projection of this point on S^n. Hence, F is a mapping of $(S^n - V) \times I$ in $S^n - V$. Clearly, $F(x, 0) = x/|x| = x$. Furthermore, $F(x, 1) = x$ for $x \in E^n_+$, and $F(x, 1) = p(x)/|p(x)| \in S^{n-1}$ for $x \in E^n_- - V$. Thus (E^n_+, S^{n-1}) is a deformation retract of $(S^n - V, E^n_- - V)$. Therefore $\epsilon:(E^n_+, S^{n-1}) \longrightarrow (S^n, E^n_-)$ is an excision, and $\epsilon_*:H_q(E^n_+, S^{n-1}) \longrightarrow H_q(S^n, E^n_-)$ is an isomorphism (even in the augmented case, since $E^n_- - V \neq \varnothing$).

The projection p of R^{n+1} upon the hyperplane defined by $x_n = 0$ yields a homeomorphism of both E^n_+ and E^n_- with E^n. Since the origin is a deformation retract of E^n, each E^n, E^n_+, and E^n_- is homotopic to a point. By Lemma 7.11, $j_*:H_q(S^n) \longrightarrow H_q(S^n, E^n_-)$ and $\partial^*:H_q(E^n_+, S^{n-1}) \longrightarrow H_{q-1}(S^{n-1})$ are both isomorphisms in the augmented case. Hence, $\partial_*\epsilon_*^{-1}j_*$ is an isomorphism of $H_q(S^n)$ on $H_{q-1}(S^{n-1})$, and therefore $H_q(S^n) \cong H_{q-n}(S^0)$. The set S_0 is the point pair ± 1 and therefore has two path-components. In the augmented theory, this means that $H_0(S^0) \cong R$ and $H(S^0) = 0$ for $q \neq 0$. Consequently, $H_n(S^n) \cong R$ and $H_q(S^n) = 0$ for $q \neq n$. The only difference between the augmented and non-augmented theories occurs when $q = 0$. For $n > 0$, the number of path-components is always one. Hence, $H_0(S^n) \cong R$ for $n > 0$ in the non-augmented case. Later, the Mayer-Vietoris Sequence will furnish a simpler way to compute $H_q(S^n)$. The results of these computations can be summarized in

7.12. *The homology groups of S^n are as follows:*

(1) *Non-augmented*

$$H_0(S^n) \cong R \text{ and } H_n(S^n) \cong R \text{ for } n > 0.$$
$$H_0(S^0) \cong R \oplus R,\ H_q(S^n) \text{ for } q \neq n, 0.$$

(2) *Augmented*

$$H_n(S^n) \cong R,\ H_q(S^n) = 0 \text{ for } q \neq n.$$

7.13. Corollary: *Spheres of different dimensions are not homotopic to each other, to E^n, or to a point.*

Proof: This follows from 7.12 and from the fact that $H_q(E^n) = 0$ for all q.

Let $f:R^{n+1} \longrightarrow R^{n+1}$ be the reflection of R^{n+1} in the hyperplane $x_0 = 0$; that is, $f(x_0, x_1, \cdots, x_n) = (-x_0, x_1, \cdots, x_n)$. For $n \geq 0$, f induces a function $f:S^n \longrightarrow S^n$.

7.14. THEOREM: *If f is the reflection in $x_0 = 0$, then in the augmented case $f_*:H_q(S^n) \longrightarrow H_q(S^n)$ is a multiplication by -1.*

Proof: Induction is employed here. For $n = 0$, S^0 consists of the two points $P = (1)$ and $Q = (-1)$. For $q \neq 0$, $H_q(S^0) = 0$. Hence, the assertion need only be proved for $f_*:H_0(S^0) \longrightarrow H_0(S^0)$. In this case the cycles have the form $r(P) - r(Q)$ with $r \in R$. Since $f(P) = Q$ and $f(Q) = P$, the sign is changed by $f^{\#}$, and hence, by f_*. The diagram

$$\begin{array}{ccc} H_q(S^{n+1}) & \xrightarrow{\partial_* \epsilon_*^{-1} j_*} & H_{q-1}(S^n) \\ \downarrow f_* & & \downarrow f_* \\ H_q(S^{n+1}) & \xrightarrow{\partial_* \epsilon_*^{-1} j_*} & H_{q-1}(S^n) \end{array}$$

is commutative. Then Lemma 7.10 yields the theorem for $q = n + 1$ whenever it is true for $q = n \geq 0$.

7.15. Lemma: *Let g be an orthogonal transformation of R^{n+1} and let* $\operatorname{Det} g = +1$. *Then the induced function $g:S^n \longrightarrow S^n$ is homotopic to the identity, and hence, $g_* = Id$.*

Proof: For $n = 0$, the only orthogonal transformations are $g = \text{Id}$ and $g = -\text{Id}$; from $\operatorname{Det} g = +1$ it follows that $g = \text{Id}$. Now suppose that the lemma has already been proved for $n - 1 \geq 0$. Let E be the plane determined by the north pole $a = (0, \cdots, 0, 1)$, $g(a) = b$, and the origin, in case those are not colinear. Otherwise let E be an arbitrary plane through the three points. Let E_0 be the orthogonal complement of E in R^{n+1}. Let h be the rotation of R^{n+1} which leaves E point-wise fixed and for which $h(b) = a$. It is easy to see that $h \sim \text{Id}$; that is, h is homotopic to the identity. Since a is fixed under hg, the subspace E' perpendicular to a is mapped onto itself by hg, and so is S^{n-1}, where $S^{n-1} = S^n \cap E'$. From $\operatorname{Det} h = \operatorname{Det} g = 1$, it follows that $\operatorname{Det} hg = 1$. The determinant of the orthogonal transformation f induced in S^{n-1} by hg is therefore also $+1$. By inductive hypothesis, $f \sim \text{Id}$, from which it follows that $hg \sim \text{Id}$, and then that $g \sim \text{Id}$.

7.16. THEOREM: *Let $g:S^n \longrightarrow S^n$ be an orthogonal trans-*

formation. Then g_* is obtained from multiplication by Det g. *(The determinant is the only homotopy invariant of the orthogonal transformations.)*

Proof: By Lemma 7.15, it follows from Det $g = +1$ that $g_* \sim \text{Id}$, and then that $g_* = \text{Id}$. Therefore, let Det $g = -1$. The function $f: S^n \longrightarrow S^n$ defined by $f(x_0, x_1, \cdots, x_n) = f(-x_0, x_1, \cdots, x_n)$ is an orthogonal transformation having determinant -1. Hence, Det $fg = +1$. Hence, $fg \sim \text{Id}$ and $f_*g_* = \text{Id}$. Since f_* is the multiplication by -1 (see Theorem 7.13), g_* is also the multiplication by -1.

As an example, consider the antipodal function $\alpha(x) = -x$. It is orthogonal; and, since Det $\alpha = (-1)^{n+1}$, α_* is the multiplication by $(-1)^{n+1}$.

A *vector field* is said to be defined on S^n if and only if to each point $x \in S^n$ a vector $a(x) \neq 0$ is associated such that the function $a(x)$ is continuous and $a(x)$ is orthogonal to x. [If $a(x)$ is localized at x, then $a(x)$ is tangential to S^n.] In this definition, S^n may not be replaced by an arbitrary homeomorphic image; only diffeomorphic images may be employed.

7.17. THEOREM:

(1) S^{2n-1} *has a vector field.*

(2) S^{2n} *has no vector field (The cowlick theorem).*

(3) *Let $a(x)$ be a vector field on S^{2n-1}. Then the function $x \longrightarrow a(x)/|a(x)|$ defines a continuous function $S^{2n-1} \longrightarrow S^{2n-1}$ which is homotopic to the identity.*

Proof:

(1) For $x = (x_0, x_1, \cdots, x_{2n-1}) \in S^{2n-1}$ let

$$a(x) = (-x_1, x_0, -x_3, x_2, \cdots, x_{2n-1}).$$

From $x \neq 0$, it follows that $a(x) \neq 0$.It is trivial that $a(x)$ is continuous and orthogonal to x.

(2) Let $a(x)$ be a vector field on S^n. Let

$$F(x, t) = x\cos(t\pi) + (a(x)/|a(x)|)\sin(t\pi)$$

for $t \in I$. Since $|F(x, t)| = 1$, F is a mapping of $S^n \times I$ into S^n. Clearly, F is continuous, $F(x, 0) = x = \text{Id } x$, $F(x, \frac{1}{2}) = a(x)/|a(x)|$, and $F(x, 1) = -x = \alpha(x)$. Therefore $\alpha \sim \text{Id}$, and $n + 1$ is even. Hence, n is odd.

(3) The function F yields the required homotopy when $0 \leq t \leq \frac{1}{2}$.

Homology Groups of Graphs

By the Direct Sum Theorem (7.2), it is sufficient to consider connected graphs X. It has already been proved that each such graph is homotopic to a line segment with a finite set of attached circles $S_1^{(i)}$, $i = 1, 2, \cdots, r$. Denote this line segment by g.

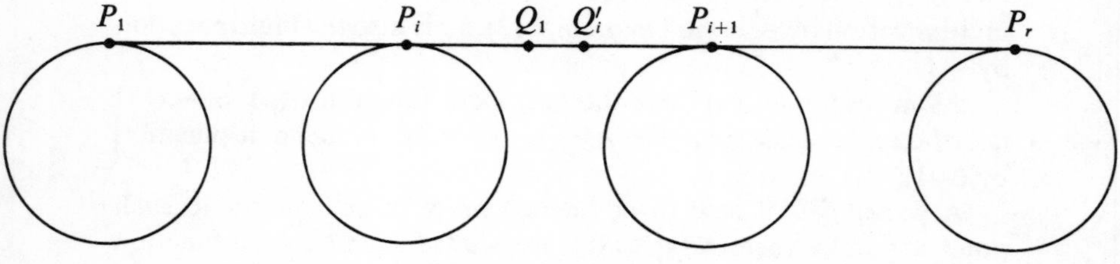

Figure 13

The augmented theory will be used here. Since g is homotopic to a point, $H_q(g) = 0$; and hence, $H_q(X) = H_q(X, g)$ by Lemma 7.8. Between P_i and P_{i+1}, choose points Q_i and Q_i'. The union of the open segments Q_iQ_i' will be denoted by V. The Excision Theorem shows that V defines an excision $(X - V, g - V) \longrightarrow (X, g)$. Now denote the union of the open segments P_iP_{i+1} by U. Since $(X - U, g - U)$ is a deformation retract of $(X - V, g - V)$, U also defines an excision by Lemma 7.7. Therefore, $H_q(X, g) \cong H_q(X - U, g - U)$. For $q \neq 0$, the Direct Sum Theorem (7.2) can now be used to yield $H_q(X - U, g - U) = \sum^{\oplus} H_q(S_1^{(i)}, P_i)$. Lemma 7.8 shows that

$$H_q(S_1^{(i)}, P_i) \cong H_q(S_1^{(i)}) = \begin{cases} R \text{ for } q = 1 \\ 0 \text{ for } q \neq 0, 1. \end{cases}$$

Since X is pathwise-connected, it can now be seen that

$$H_q(X) \cong \begin{cases} R \oplus R \oplus \cdots \oplus R \ (r \textit{ summands}) \textit{ for } q = 1 \\ 0 \qquad \textit{for } q \neq 0, q \neq 1. \end{cases}$$

7.18.

$$H_0(X) \cong \begin{cases} 0 \textit{ in the augmented theory} \\ R \textit{ in the non-augmented theory.} \end{cases}$$

If R is chosen to be a field, then $H_1(X)$ is a vector space of dimension r. Thus, the number of circles is a homology invariant and indeed, it is the only one.

In the theory of function of a complex variable, there often occur plane domains which are bounded by finitely many simply connected (usually piecewise smooth) curves as in the following figure:

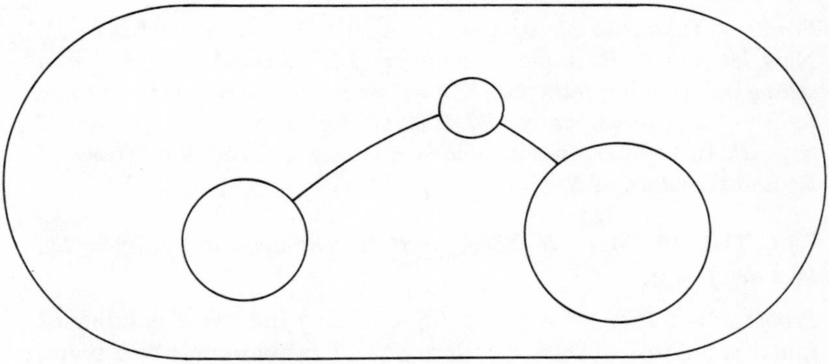

Figure 14

If the boundary curves are joined as in Figure 14, then it can easily be seen that the domain is homotopic to a graph that consists of the inner boundary curves and the connecting paths. The homology group of such a domain is therefore known by 7.17. A domain of this type is called *n-tuply connected* if and only if $H_1(X)$ has $n - 1$ direct summands R. It should be noted that the term *n-connected* is reserved for another concept.

Degree of a Function $f: S^n \longrightarrow S^n$

Let $f:S^n \longrightarrow S^n$ be continuous. Let R be the ring of integers. Then $H_n(S^n) = Z$, and $H_n(S^n)$ is a free Z-module with one generator c. For the homomorphism $f_*:H_n(S^n) \longrightarrow H_n(S^n)$, it follows that $f_*(c) = mc$ for some $m \in Z$. Then $f_*(rc) = rf_*(c) = rmc$. In other words, f_* is a multiplication by the fixed integer m. This integer m is called the *degree of f*. Obviously, $\deg fg = \deg f \cdot \deg g$. Theorem 7.16 can therefore be written: *The degree of an orthogonal transformation $f:S^n \longrightarrow S^n$ is Det f.* For a homeomorphism $f:S^n \longrightarrow S^n$,

$$\deg f \cdot \deg f^{-1} = \deg \mathrm{Id} = 1.$$

Hence, $\deg f = \pm 1$. Obviously, homotopic functions have the same degree. By a theorem of H. Hopf, which will not be proved here, it follows from $\deg h = \deg f$ that $h \sim f$.

7.18. Lemma: *Let P be a point of S^n. Then $S^n - P$ is homotopic to a point.*

Proof: Take P to be $(0, 0, \cdots, -1)$. For $n = 0$, $S^0 - P$ is $\{+1\}$. Now let $n \geq 1$. Example 8 following 5.25 showed that S^{n-1} is a strong deformation retract of $E^n_- - P$ since $E^n_- - P$ is homeomorphic to $E^n - 0$. Consequently, E^n_+ is a strong deformation retract of $S^n - P$. In turn, the north pole is a strong deformation retract of E^n_+ and therefore of $S^n - P$.

7.19. THEOREM: *If $f:S^n \longrightarrow S^n$ is continuous and $f(S^n) \neq S^n$, then $\deg f = 0$.*

Proof: Let $P \in S^n$ and $P \notin f(S^n)$. Then f induces a continuous function $g:S^n \longrightarrow (S^n - P)$. Since $S^n - P$ is homotopic to a point, $H_n(S^n_n - P) = 0$, and the homomorphism $g_*:H_n(S^n) \longrightarrow H_n(S^n - P)$ is the zero homomorphism. Let $i:(S^n - P) \longrightarrow S^n$ be the injection. Then $f = ig$ and $f_* = i_* g_* = 0$. Thus, $\deg f = 0$.

7.20. Lemma: *Let $n \geq 1$, let E^n be the closed ball $|x| \leq 1$ in R^n, and let S^{n-1} be its boundary. Let $f:E^n \longrightarrow S^{n-1}$ be continuous and let g be the restriction of f to S^{n-1}. Then g has degree zero.*

Proof: Define $F(x, t):S^{n-1} \times I \longrightarrow S^{n-1}$ by $F(x, t) = f(tx)$. Then $F(x, 1) = f(x) = g(x)$ for $x \in S^{n-1}$, and $F(x, 0) = f(0)$. Hence, g is homotopic to the function $x \longrightarrow f(0)$, which has degree 0 by Theorem 7.19. Therefore, $\deg g = 0$.

7.21. Corollary: *S^{n-1} is not a retract of E^n.*

Proof: If $f:E^n \longrightarrow S^{n-1}$ is a retraction, then $f(x) = x$ when $x \in S^{n-1}$. Hence, $g = \mathrm{Id}$ and $g_* = 0 = \mathrm{Id}$. This contradicts $H_n(S^n) \neq 0$.

There are other ways of proving 7.21; for instance,

7.22. THEOREM: *Let $Y \in X$ be a retract of X and let $i: Y \longrightarrow X$ be the injection. Then $i_*:H_q(Y) \longrightarrow H_q(X)$ is a monomorphism.*

Proof: For a retraction $f\colon Y \longrightarrow X$, it is true that $fi = \text{Id}$ and then $f_* i_* = \text{Id}$. Hence, i_* is a monomorphism since it follows from $i_*(c) = 0$ that $f_* i_*(c) = 0$ and $c = 0$.

7.23. Corollary: *If there is a q for which $H_q(Y) \neq 0$ but $H_q(X) = 0$, Y is not a retract of X.*

Special choices of X and Y yield the following results:

(a) S^{n-1} is not a retract of E^n since $H_{n-1}(S^{n-1}) \neq 0$ and $H_{n-1}(E^n) = 0$.

(b) If S_1^r is a subspace of S^n homeomorphic to S^r for $r < n$, S_1^r is not a retract of S^n since $H_r(S_1^r) \neq 0$ and $H_r(S^n) = 0$.

7.24. THE BROUWER FIXED-POINT THEOREM: *If $f\colon E^n \longrightarrow E^n$ is continuous, then f has a fixed point.*

Proof: There is nothing to be proved for $n = 0$ since E^0 is a point. Let $n \geq 1$. Suppose that $f(x) \neq x$ for all $x \in E^n$. For $x \in E^n$, define $g(x)$ as the intersection of S^{n-1} with the open half-line which starts at $f(x)$ and contains x. An equation for this half-line is $w = f(x) + t(x - f(x))$, $t > 0$. The continuity of g will now be proved. Let $f(x) = b$ and $x - f(x) = a$. Then, if $w \in S^{n-1}$, $w^2 = a^2 t_0^2 + 2t_0(ab) + b^2 = 1$, where ab denotes the scalar product. Since $t_0 > 0$, the only solution under consideration is

$$t_0 = \frac{1}{a^2}\left(-ab + \sqrt{(ab)^2 + a^2(1 - b)^2}\right).$$

This root is real, because $b^2 \leq 1$. For $b^2 < 1$, it follows that $t_0 > 0$. For $b^2 = 1$, it follows from $x - b = a$ that

$$ab = xb - b^2 = xb - 1,$$

and then from $xb \leq |x| \cdot |b| \leq 1$ that $xb - 1 \leq 0$. From $xb = 1$, it follows that $x^2 = b^2 = 1$, and therefore that

$$(x - b)^2 = x^2 - 2xb + b^2 = 0,$$

but then $x = b = f(x)$ contrary to assumption. Therefore, $t_0 > 0$ even when $b^2 = 1$. This has proved the existence of a unique intersection point of the half-line with S^{n-1}. The formula shows that the intersection is a continuous function of a and b, and hence, of x. For $x \in S^{n-1}$, the intersection point is $g(x) = x$. Hence, g is a retraction of E^n on S^{n-1}. This contradicts 7.21. Therefore, f has a fixed point.

Remark: Proofs of the Brouwer Fixed-Point Theorem in analysis often involve stronger assumptions about f.

7.25. THEOREM: *If f and g are continuous functions on S^n into S^n and $f(x) \neq g(x)$ for all $x \in S^n$, then* $\deg f = (-1)^{n+1} \deg g$.

Proof: Let

$$F(x, t) = \frac{tf(x) - (1 - t)g(x)}{|tf(x) - (1 - t)g(x)|} \qquad \text{for } x \in S^n, 0 \leq t \leq 1.$$

The denominator is non-zero for all values of x and t, for it follows from $|f(x)| = |g(x)| = 1$ and $tf(x) = (1 - t)g(x)$ that $t = 1 - t$ and $f(x) = g(x)$. Hence, $F(x, t)\colon S^n \times I \longrightarrow S^n$ is continuous. Since $F(x, 0) = -g(x) = \alpha \cdot g(x)$ and $F(x, 1) = f(x)$, the functions f and αg are homotopic. Therefore, $\deg f = \deg \alpha g = \deg \alpha \cdot \deg g = (-1)^{n+1} \deg g$.

7.26. Corollary: *If $f\colon S^n \longrightarrow S^n$ has no fixed points, then $\deg f = (-1)^{n+1}$.*

Proof: Let $g = \text{Id}$ in 7.25.

An example of a fixed-point free function is the antipodal map α.

7.27. Corollary: *If $f\colon S^n \longrightarrow S^n$ is continuous and $f(S^n) \neq S^n$, then f has a fixed point, and there is a point which is mapped by f on its antipodal point.*

Proof: By Theorem 7.19, $\deg f = 0$. If β is an orthogonal transformation, then $\deg \beta = \pm 1 \neq (-1)^{n+1} \deg f$. By Theorem 7.25, there is an $x \in S^n$ for which $f(x) = \beta(x)$. The choices $\beta = \text{Id}$ and $\beta = \alpha$ yield an $x \in S^n$ for which $f(x) = x$ and a $y \in S_n$ for which $f(y) = -y$, respectively.

8

The Tensor Product

Let R be a commutative ring with identity. Let $A_1, A_2, \cdots, A_n$ and C be unitary R-modules. Let $\prod_i A_i$ be the cartesian product of the A_i.

Definition: A function $f: \prod_i A_i \longrightarrow C$ is *multilinear* if and only if for each j, $1 \leq j \leq n$, and each choice of $a_1, a_2, \cdots, a_{j-1}, a_{j+1}, \cdots, a_n$, where $a_i \in A_i$, the function f is an R-linear function sending A_j into C; that is,

$$f(\cdots, a_j + b_j, \cdots) = f(\cdots, a_j, \cdots) + f(\cdots, b_j, \cdots)$$

and

$$f(\cdots, ra_j, \cdots) = rf(\cdots, a_j, \cdots)$$

for $a_j, b_j \in A_j$ and $r \in R$. In the case $n = 2$, a multilinear function $f{:}A_1 \times A_2 \longrightarrow C$ is also termed bilinear. If f is interpreted as a product, then bilinearity means that the product is distributive.

The image of a multilinear function does not need to be an R-module. The multilinear function f is called *surjective* if and only if $\operatorname{Im} f$ generates C. In this case, the relation

$$rf(a_1, a_2, \cdots, a_n) = f(ra_1, a_2, \cdots, a_n)$$

shows that each $c \in C$ is the finite sum of images $f(a_1, a_2, \cdots, a_n)$.

Definition: By a *tensor product* of the modules A_i, $i = 1, 2, \cdots, n$ is meant an R-module M together with a multilinear function $F{:}\ \prod_i A_i \longrightarrow M$ with the following properties:

(1) F is surjective.

(2) To each multilinear mapping $f{:}\prod_i A_i \longrightarrow C$, there is a homomorphism $\tilde{f}{:}M \longrightarrow C$ with $f = \tilde{f}F$. In other words, the diagram

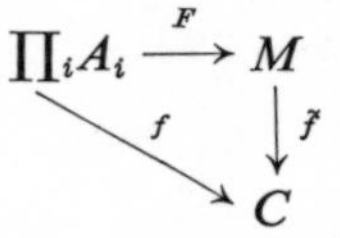

is commutative. Obviously, f is only surjective if the homomorphism $\tilde{f}$ is surjective in the usual sense; that is, if $\tilde{f}$ is an epimorphism. If each A_i is replaced with an isomorphic B_i, then M is also the tensor product of the B_i.

8.1. THEOREM: *Let M_1, F_1 and M_2, F_2 be two tensor products of the modules A_i, $i = 1, 2, \cdots, n$. Then $\tilde{F}_1{:}M_2 \longrightarrow M_1$ and $\tilde{F}_2{:}\ M_1 \longrightarrow M_2$ are mutually reciprocal isomorphisms.*

Proof: $F_1 = \tilde{F}_1F_2$ and $F_2 = \tilde{F}_2F_1$, and therefore $F_1 = \tilde{F}_1\tilde{F}_2F_1$. For each $c \in F_1(a_1, \cdots, a_n)$, it therefore follows that $\tilde{F}_1\tilde{F}_2c = c$. Since M is generated by such elements c, it follows that $\tilde{F}_1\tilde{F}_2 = \mathrm{Id}$. Similarly, $\tilde{F}_2\tilde{F}_1 = \mathrm{Id}$.

Theorem 8.1 shows that a tensor product of the A_i, $i = 1, \cdots, n$ is unique up to an isomorphism. For the sake of convenience, the proof of the existence of such a product will be postponed. Usually one writes $A_1 \otimes \cdots \otimes A_n = \otimes_{i=1}^n A_i$ instead of M and $a_1 \otimes \cdots \otimes a_n$ instead of $F(a_1, \cdots, a_n)$. The formula $f = \tilde{f}F$ then becomes

$f(a_1, \cdots, a_n) = \tilde{f}(a_1 \otimes \cdots \otimes a_n)$. Since $\tilde{f}: M \longrightarrow C$ is a homomorphism, it follows that

$$\tilde{f}(\sum a_1 \otimes \cdots \otimes a_n) = \sum f(a_1, \cdots, a_n).$$

8.2. THEOREM (The Associative Law): *If* $A = \otimes_{i=1}^{n} A_i$, $B = \otimes_{j=1}^{m} B_j$, *and* $A \otimes B$ *exists, then the function*

$$F: \prod_i A_i \times \prod_j B_j \longrightarrow A \otimes B$$

which is defined by

$$F(a_1, \cdots, a_n, b_1, \cdots, b_m) = (a_1 \otimes \cdots \otimes a_n) \otimes (b_1 \otimes \cdots \otimes b_m)$$

yields a tensor product of $A_1, \cdots, A_n, B_1, \cdots, B_m$.

Proof:

(1) The elements of $A \otimes B$ are finite sums $\sum a \otimes b$, where $a \in A$ and $b \in B$. Here,

$$a = \sum a_1 \otimes \cdots \otimes a_n \text{ and } b = \sum b_1 \otimes \cdots \otimes b_m.$$

Multilinearity then yields

$$\begin{aligned} \sum a \otimes b &= \sum (a_1 \otimes \cdots \otimes a_n) \otimes (b_1 \otimes \cdots \otimes b_m) \\ &= \sum F(a_1, \cdots, a_n, b_1, \cdots, b_m). \end{aligned}$$

Therefore, F is surjective.

(2) Now let $f: \prod_i A_i \times \prod_j B_j \longrightarrow C$ be an arbitrary multilinear function. The problem is to find a suitable homomorphism of $A \otimes B$ on C. For $b_1, \cdots, b_m$ fixed, $f(a_1, \cdots, a_n, b_1, \cdots, b_n)$ yields a multilinear function of $\prod_i A_i$ into C. Therefore, there is a homomorphism $\tilde{f}_{b_1 \cdots b_m}: A \longrightarrow C$ with

$$\tilde{f}_{b_1 \cdots b_m}(a_1 \otimes \cdots \otimes a_n) = f(a_1, \cdots, a_n, b_1, \cdots, b_m).$$

For $a = \sum a_1 \otimes \cdots \otimes a_n$, it follows that

$$\tilde{f}_{b_1 \cdots b_m}(a) = \sum f(a_1, \cdots, a_n, b_1, \cdots, b_m).$$

For a fixed a, the right-hand side indicates that $\tilde{f}_{b_1 \cdots b_m}(a)$ is a multilinear function sending $\prod_j B_j$ into C. Therefore, there is a homomorphism $\tilde{\tilde{f}}_a: B \longrightarrow C$ such that

$$\tilde{\tilde{f}}_a(b_1 \otimes \cdots \otimes b_m) = \tilde{f}_{b_1 \cdots b_m}(a)$$
$$= \sum f(a_1, \cdots, a_n, b_1, \cdots, b_m).$$

When $b = \sum b_1 \otimes \cdots \otimes b_m \in B$, it follows that

$$\tilde{\tilde{f}}_a(b) = \sum f(a_1, \cdots, a_n, b_1, \cdots, b_m).$$

The function $A \times B \longrightarrow C$ for which $(a, b) \longrightarrow \tilde{\tilde{f}}_a(b)$ is trivially bilinear. Hence, there is a homomorphism $\tilde{f}: A \otimes B \longrightarrow C$ for which

$$\tilde{f}(a_1 \otimes \cdots \otimes a_n \otimes b_1 \otimes \cdots \otimes b_m)$$
$$= f(a_1, \cdots, a_n, b_1 \cdots b_m).$$

Therefore, $A \otimes B = (A_1 \otimes \cdots \otimes A_n) \otimes (B_1 \otimes \cdots \otimes B_m)$ is a tensor product of $A_1, \cdots, A_n, B_1, \cdots, B_m$.

8.3. Corollary: *If the tensor product of any two factors exists, then the tensor product of n factors exists. Furthermore, the associative law holds up to an isomorphism.*

8.4. THEOREM (The Commutative Law): *If $A \otimes B$ exists, then it can be employed as the tensor product $B \otimes A$ by means of the function $F(b, a) = a \otimes b$.*

Proof:

(1) F is surjective, since $A \otimes B$ is generated by the terms $a \otimes b$.

(2) Suppose that $f: B \times A \longrightarrow C$ is bilinear. The same f can be interpreted as a bilinear map of $A \times B$ into C. Therefore, there is a homomorphism $\tilde{f}: A \otimes B \longrightarrow C$ such that $\tilde{f}(a \otimes b) f(b, a)$.

Warning! It has only been proved that, with the proper F, $A \otimes B$ can be interpreted as $B \otimes A$, or in other words, that $A \otimes B \cong B \otimes A$. Even in the case $B = A$, it usually happens that $a_1 \otimes a_2 \neq a_2 \otimes a_1$ for a_1 and a_2 in A. Otherwise, $f(a_1, a_2) = f(a_2, a_1)$ for each bilinear mapping f, and this is false.

8.5. THEOREM: *Let $A = \sum_i^{\oplus} A_i$ be a direct sum of an arbitrary collection of R-modules A_i. If the tensor product $A_i \otimes B$ exists for the R-module B and each A_i, then $M = \sum_i^{\oplus}(A_i \otimes B)$ is a tensor product $A \otimes B$ provided that $a \otimes b$ is defined to be $\sum_i (a_i \otimes b)$ for $a = \sum_i a_i \in A$ and $b \in B$.*

Proof:

(1) Since the sum is direct, the term $a \otimes b$ is well defined. The function is obviously multilinear. The elements $a_i \otimes b$ generate $A_i \otimes B$ and consequently, also M.

(2) Let $f{:}A \times B \longrightarrow C$ be multilinear. A function $\bar{f}$ must be found. Let f_i be the restriction of f to $A_i \times B$. To the bilinear f_i there is a homomorphism $\bar{f}_i{:}A_i \otimes B \longrightarrow C$ such that

$$\bar{f}_i(a_i \otimes b) = f_i(a_i, b) = f(a_i, b).$$

For $x \in M$, where $x = \sum x_i$ and $x_i \in A_i \otimes B$, define $\bar{f}(x)$ by $\sum_i \bar{f}_i(x_i)$. This defines a homomorphism $\bar{f}{:}M \longrightarrow C$. When $a = \sum a_i \in A$, then $a \otimes b = \sum (a_i \otimes b)$ and therefore,

$$\begin{aligned}\bar{f}(a \otimes b) &= \bar{f}(\textstyle\sum_i (a_i \otimes b)) \\ &= \textstyle\sum_i \bar{f}_i(a_i \otimes b) = \sum_i f(a_i, b) = f(a, b).\end{aligned}$$

8.6. Corollary: *By commutativity, Theorem 8.5 also holds for direct decompositions of the second factor. By associativity, it holds for a finite number of factors.*

8.7. THEOREM: *Each R-module B is also a tensor product $R \otimes B$ if the definition $r \otimes b = rb$ is made for $r \in R$ and $b \in B$.*

Proof:

(1) The function $\otimes{:}R \times B \longrightarrow B$ is well defined, bilinear and, as $r = 1$ shows, surjective.

(2) Let $f{:}R \times B \longrightarrow C$ be bilinear. Define $\bar{f}{:}B \longrightarrow C$ by $\bar{f}(b) = f(1, b)$. Then $\bar{f}$ is a homomorphism, and

$$f(r, b) = rf(1, b) = \bar{f}(rb) = \bar{f}(r \otimes b).$$

8.8. Corollary: *By commutativity, each R-module B is a tensor product $B \otimes R$. By Theorem 8.5, the product $A \otimes B$ exists when one of the factors is a free module (the direct sum of modules isomorphic to R). If A is free, then $A \otimes B$ is the direct sum of just as many modules isomorphic to B as there are basis elements in A; this number need not be finite.*

8.9. THEOREM: *Suppose that $B \otimes M$ exists. Let A be a submodule of B and let H be the submodule of $B \otimes M$ generated by all*

$a \otimes m$, where $a \in A$ and $m \in M$. Then the factor module $(B \otimes M)/H$ becomes a tensor product $(B/A) \otimes M$ by means of the definition

8.10. $$(b + A) \otimes m = (b \otimes m) + H.$$

Remark: In Example (4) following Corollary 8.17, it will be proved that in general $H \neq A \otimes M$. The restriction of $\otimes: B \times M \longrightarrow B \times M$ to $A \times M$ yields only an epimorphism of $A \otimes M$ on H. In case A is a direct summand of B, then $H = A \otimes M$.

Proof of Theorem 8.9: The homomorphism of $(B/A) \times M$ into $(B \otimes M)/H$ given by 8.10 is well defined and trivially bilinear. It is surjective because $(B \otimes M)/H$ is generated by the images $(b \otimes m) + H$. Now let $f:(B/A) \times M \longrightarrow C$ be bilinear. Define $F:B \times M \longrightarrow C$ by $F(b, m) = f(b + A, m)$. To the bilinear function F there corresponds a homomorphism $\tilde{F}:B \otimes M \longrightarrow C$ such that

$$\tilde{F}(b \otimes m) = F(b, m) = f(b + A, m).$$

For $a \in A$, this means that

$$\tilde{F}(a \otimes m) = f(a + A, m) = 0.$$

Hence $\tilde{F}$ vanishes on H, and induces, according to Lemma 1.1, a homomorphism $\tilde{f}(B \otimes M)/H \longrightarrow C$ by means of $\tilde{f}(x + H) = \tilde{F}(x)$ for $x \in B \otimes M$. Clearly,

$$f(b \otimes m + H) = \tilde{F}(b \otimes m) = f(b + A, m).$$

Then 8.10 yields

$$\tilde{f}((b + A) \otimes m) = f(b + A, m).$$

8.11. Lemma: *Each module is isomorphic to a factor module of a free module.*

Proof: Let x_i, $i \in I$ be a generating system of the module N. Then $N = \sum Rx_i$, where the sum does not need to be direct. Form the free module $L = \sum^{\oplus} Ry_i$. Define an epimorphism of L on N by $y_i \longrightarrow x_i$. Then $N = L/A$ where A is the kernel.

8.12. THEOREM (Existence of the Tensor Product): *For each two R-modules N and M, there exists a tensor product $N \otimes M$.*

Proof: N is isomorphic to L/A, where L is a free module. By Theorems 8.6 and 8.7, the product $L \otimes M$ exists. By Theorem 8.9, $(L/A) \otimes M$ exists and is a tensor product $N \otimes M$.

This completes the proof of the existence of the tensor product of a finite set of R-modules (see Corollary 8.3).

Tensor Products of Functions

8.13. THEOREM: *Let $f:A \longrightarrow A'$ and $g:B \longrightarrow B'$ be homomorphisms. Then there is a homomorphism $f \otimes g: A \otimes B \longrightarrow A' \otimes B'$ which is defined by $(f \otimes g)(a \otimes b) = f(a) \otimes f(b)$ and which is called the tensor product of f and g.*

Proof: Define $h:A \times B \longrightarrow A' \otimes B'$ by $h(a, b) = f(a) \otimes g(b)$. Since $\otimes$ is bilinear, h is also bilinear. Consequently, there is a homomorphism $\tilde{h}:A \otimes B \longrightarrow A' \otimes B'$ such that $\tilde{h}(a \otimes b) = f(a) \otimes g(b)$. Let $f \otimes g = \tilde{h}$.

The following computational rules follow easily by applying the relevant functions to $a \otimes b$:

(a) For $f, f_1:A \longrightarrow A'$ and $g, g_1:B \longrightarrow B'$, both

$$(f + f_1) \otimes g = f \otimes g + f_1 \otimes g$$

and

$$f \otimes (g + g_1) = f \otimes g + f \otimes g_1.$$

(b) For $A \xrightarrow{f} A' \xrightarrow{f'} A''$ and $B \xrightarrow{g} B' \xrightarrow{g'} B''$,

$$(f' \otimes g')(f \otimes g) = (f'f) \otimes (g'g).$$

(c) For $r \in R$,

$$r(f \otimes g) = (rf) \otimes g = f \otimes rg.$$

Remark: The function $(f, g) \longrightarrow f \otimes g$ is a bilinear function

$$\mathrm{Hom}(A, A') \times \mathrm{Hom}(B, B') \longrightarrow \mathrm{Hom}(A \otimes B, A' \otimes B').$$

However, the induced homomorphism

$$\mathrm{Hom}(A, A') \otimes \mathrm{Hom}(B, B') \longrightarrow \mathrm{Hom}(A \otimes B, A' \otimes B')$$

is not necessarily either a monomorphism or an epimorphism.

8.14. THEOREM: *Let*

$$A \xrightarrow{i} B \xrightarrow{j} C \longrightarrow 0$$

be exact and let M be an R-module. Then

$$A \otimes M \xrightarrow{i \otimes 1} B \otimes M \xrightarrow{j \otimes 1} C \otimes M \longrightarrow 0$$

is exact.

Proof: Since j is an epimorphism, $(j \otimes 1)(b \otimes m) = jb \otimes m$ shows that $j \otimes 1$ is an epimorphism. Furthermore,

$$(j \otimes 1)(i \otimes 1) = (ji) \otimes 1 = 0 \otimes 1 = 0.$$

Hence, it remains to be proved that to $h \in \text{Ker}(j \otimes 1)$ there is an $x \in A \otimes M$ such that $(i \otimes 1)x = h$. Since $\text{Ker} j = iA$, j induces an isomorphism $\tilde{j}: B/iA \longrightarrow C$ by means of $\tilde{j}(b + iA) = jb$. Let H be the submodule of $B \otimes M$ generated by the set of all $i(a) \otimes m$, $a \in A$. By Theorem 8.9, $(B \otimes M)/H$ is a tensor product of B/iA and M under the definition $(b + iA) \otimes m = b \otimes m + H$. Since $\tilde{j}^{-1}$ exists, the isomorphisms $\tilde{j} \otimes 1: (B \otimes M) / H \longrightarrow C \otimes M$ and $\tilde{j}^{-1} \otimes 1: C \otimes M \longrightarrow (B \otimes M)/H$ are mutually reciprocal. For $\tilde{j} \otimes 1$, the definitions show that

$$\begin{aligned}(\tilde{j} \otimes 1)(b \otimes m + H) &= (\tilde{j} \otimes 1)((b + iA) \otimes m) \\ &= \tilde{j}(b + iA) \otimes m = jb \otimes m.\end{aligned}$$

For $j \otimes 1: B \otimes M \longrightarrow C \otimes M$, the definition yields

$$(j \otimes 1)(b \otimes m) = jb \otimes m.$$

Since $\tilde{j} \otimes 1$ is an isomorphism, it follows that $\text{Ker}(j \otimes 1) = H$. If $h \in \text{Ker}(j \otimes 1)$, it can be written as

$$h = \sum i(a) \otimes m = \sum (i \otimes 1)(a \otimes m).$$

Choose $x = \sum a \otimes m$ to find the desired element.

8.15. THEOREM: *Let* $A \xrightarrow{i} B \xrightarrow{j} C \longrightarrow 0$ *and* $A' \xrightarrow{i'} B' \longrightarrow C' \xrightarrow{j'} 0$ *be exact. Then the "nine-diagram"*

$$\begin{array}{ccccccc}
A \otimes A' & \xrightarrow{i\otimes 1} & B \otimes A' & \xrightarrow{j\otimes 1} & C \otimes A' & \longrightarrow & 0 \\
\downarrow & & \downarrow & & \downarrow & & \\
A \otimes B' & \xrightarrow{i\otimes 1} & B \otimes B' & \xrightarrow{j\otimes 1} & C \otimes B' & \longrightarrow & 0 \\
\downarrow & & \downarrow & & \downarrow & & \\
A \otimes C' & \xrightarrow{i\otimes 1} & B \otimes C' & \xrightarrow{j\otimes 1} & C \otimes C' & \longrightarrow & 0 \\
\downarrow & & \downarrow & & \downarrow & & \\
0 & & 0 & & 0 & &
\end{array}$$

is exact and each rectangle commutes. Furthermore,

8.16. $(A \otimes B') \oplus (B \otimes A') \xrightarrow{(i\otimes 1) + (1\otimes i')} B \otimes B' \xrightarrow{j\otimes j'} C \otimes C' \longrightarrow 0$
is also exact.

Proof: Theorem 8.14 shows the exactness of the "nine-diagram." The commutativity follows from

$$(f \otimes 1)(1 \otimes g) = f \otimes g = (1 \otimes g)(f \otimes 1).$$

The exactness of 8.16 remains to be proved. For this purpose, consider the diagram,

$$\begin{array}{ccccccc}
 & & A_2 & \xrightarrow{\phi} & A_3 & \longrightarrow & 0 \\
 & & \downarrow{\scriptstyle \alpha} & & \downarrow{\scriptstyle \beta} & & \\
B_1 & \xrightarrow{\chi} & B_2 & \xrightarrow{\psi} & B_3 & \longrightarrow & 0 \\
 & & \downarrow{\scriptstyle \gamma} & & \downarrow{\scriptstyle \delta} & & \\
 & & C_2 & \xrightarrow{\rho} & C_3 & & \\
 & & & & \downarrow & & \\
 & & & & 0 & &
\end{array}$$

whose rows and columns are exact and whose two rectangles commute. It needs to be checked that

$$B_1 \oplus A_2 \xrightarrow{\chi\oplus\alpha} B_2 \xrightarrow{\delta\psi} C_3 \longrightarrow 0$$

is exact: Since ψ and δ are epimorphisms, $\delta\psi$ is also. Furthermore,

$$\delta\psi(\chi \oplus \alpha) = \delta\psi\chi \oplus \delta\psi\alpha = 0 \oplus \rho\gamma\alpha = 0 \oplus 0 = 0.$$

Finally, let $b_2 \in B_2$ and $\delta\psi(b_2) = 0$. Because $\delta(\psi(b_2)) = 0$, there is an $a_3 \in A_3$ for which $\beta a_3 = \psi(b_2)$. To a_3 there is an a_2 for which $\phi(a_2) = a_3$. Hence, $\psi(b_2) = \beta\phi(a_2) = \psi\alpha(a_2)$. Since $\psi(b_2 - \alpha(a_2)) = 0$, there is a $b_1 \in B_1$ for which $\chi(b_1) = b_2 - \alpha(a_2)$. This shows that b_2 is the image of (b_1, a_2) under $\chi \oplus a$.

8.17. Corollary: *Let A be a submodule of B and A' a submodule of B'. Let $i: A \longrightarrow B$ and $i': A' \longrightarrow B'$ be the injections, and let H be the submodule*

$$H = ((i \otimes 1) \oplus (1 \otimes i'))((A \otimes B') \oplus (B \otimes A'))$$

of $B \otimes B'$.
Then,

$$(B/A) \otimes (B'/A') \cong (B \otimes B')/H.$$

Proof: For the exact sequences

$$A \xrightarrow{i} B \longrightarrow B/A \longrightarrow 0$$

and

$$A' \xrightarrow{i'} B' \longrightarrow B'/A' \longrightarrow 0,$$

the exactness of sequence 8.16 provides the exactness of

$$(A \otimes B') \oplus (B \otimes A') \longrightarrow B \otimes B' \longrightarrow (B/A) \otimes (B'/A') \longrightarrow 0.$$

This means that

$$(B/A) \otimes (B'/A') \cong (B \otimes B')/H.$$

Hence H is the submodule of $B \otimes B'$ generated by all $a \otimes b'$, $a \in A$, $b' \in B$ and all $b \otimes a'$, $b \in B$, $a' \in A'$. In the case $A' = 0$ and $B' = M$, the lemma reduces to Theorem 8.9. It should be noticed again that $H = (i \otimes 1)(A \otimes M)$ is not in general isomorphic to $A \otimes M$.

Examples:

(1) Let A be an R-module and $\mathscr{A}$ be an ideal, hence a submodule of R. By Theorem 8.7, $R \otimes A \cong A$. By Theorem 8.9,

$$R/(\mathscr{A} \otimes A) \cong (R \otimes A)/H \cong A/H$$

in which H is the submodule generated by the set of all elements $\alpha \otimes a = \alpha a$ where $\alpha \in \mathscr{A}$ and $a \in A$. Therefore,

$$R/(\mathscr{A} \otimes A) \cong A/\mathscr{A}A.$$

(2) Let $\mathscr{A}$ and $\mathscr{B}$ be ideals in R. By Corollary 8.17,

$$R/\mathscr{A} \otimes R/\mathscr{B} \cong R/H$$

where

$$H = ((i \otimes 1) \oplus (1 \otimes i'))(\mathscr{A} \otimes R) \oplus (R \otimes b)$$
$$= \mathscr{A}R + \mathscr{B}R = \mathscr{A} + \mathscr{B}.$$

It follows that

$$R/\mathscr{A} \otimes R/\mathscr{B} = R/(\mathscr{A} + \mathscr{B}).$$

For relatively prime ideals $\mathscr{A}$ and $\mathscr{B}$, it is true that $\mathscr{A} + \mathscr{B} = R$. Hence, $R/\mathscr{A} \otimes R/\mathscr{B} = 0$. Consider, for instance, the case $R = Z, \mathscr{A} = (2)$ and $\mathscr{B} = (3)$. Then $(Z/(2)) \otimes (Z/(3)) = 0$. The vanishing of $A \otimes B$ means that there are no non-trivial bilinear functions on $A \times B$.

(3) Let R be a principal ideal ring. By the main theorem for abelian groups, each finitely generated R-module has the form of a finite direct sum $\sum_i^{\oplus} R/(\alpha_i)$. Theorem 8.5 implies that

$$(\textstyle\sum_i^{\oplus} R/(\alpha_i)) \otimes (\sum_j^{\oplus} R/(\beta_j)) = \sum_{i,j}^{\oplus} R/(\delta_{ij})$$

where $(\delta_{ij}) = (\alpha_i) + (\beta_j)$ in accordance with Example 2. This yields a survey of all tensor products of finitely generated modules over a principal ideal ring R. No such survey is known unless R is a principal ideal ring.

(4) Consider the special case where $R = Z$, $B = Z/(4)$, $A = 2Z/(4) \subset B$, and $M = Z/(2)$ are substituted in Example (3). Then $A = Z/(2)$; and, consequently, $A \otimes M = (Z/(2)) \otimes (Z/(2)) = (Z/(2))$. Let H be the submodule of $B \otimes M$ generated by all $a \otimes m$, where $a \in A$ and $m \in M$. Since $a = 2b_1$, where $b \in Z$, it follows that $a \otimes m = 2b \otimes m = b \otimes 2m = 0$ and therefore, that $H = 0$. Thus, $A \otimes M \neq H$.

Definition: Let R be an integral domain. An R-module M is called *divisible* if and only if to each $m \in M$ and $r \in R, r \neq 0$, there exists an $m' \in M$ for which $m = rm'$. Here, m' is not necessarily uniquely defined by r and m. An R-module A is called a *torsion module* if and only if to each $a \in A$ there is an $r \in R$, $r \neq 0$, for which $ra = 0$.

8.18. THEOREM: *Let M be divisible and let A be a torsion module. Then, $A \otimes M = 0$.*

Proof: Let $a \in A$ and $m \in M$. Choose $r \in R$ such that $r \neq 0$ and $ra = 0$. Find $m' \in M$ for which $m = rm'$. Then it follows that

$$a \otimes m = a \otimes (rm') = (ra) \otimes m = 0 \otimes m = 0.$$

As an example, let Q be the Z-module of rational numbers. Since, for $0 \leq p/q < 1$ and any integer $n \neq 0$, it is clear that $p/q = n(p/nq)$, it follows that Q/Z is a divisible Z-module. Furthermore, since $q(P/q) \equiv 0$, Q/Z is a torsion module. Therefore, $(Q/Z) \otimes_Z (Q/Z) = 0$. Here, the notation $\otimes_R$ indicates that the tensor product depends on the ring. That this is so in the case at hand can be seen from

$$(Q/Z) \otimes_{Q/Z} (Q/Z) = Q/Z \neq 0.$$

Exercise: Show that $Q \otimes_Z Q = Q$ under the definition

$$a \otimes b = ab \qquad \text{for } a, b \in Q.$$

8.19. THEOREM: *Let F be a free R-module. From the exactness of*

$$0 \longrightarrow A \overset{i}{\longrightarrow} B \overset{j}{\longrightarrow} C \longrightarrow 0$$

follows the exactness of

$$0 \longrightarrow A \otimes F \overset{i \otimes 1}{\longrightarrow} B \otimes F \overset{j \otimes 1}{\longrightarrow} C \otimes F \longrightarrow 0.$$

If $i(A)$ is a direct summand of B, then $(i \otimes 1)(A \otimes F)$ is also a direct summand of $B \otimes F$. (Hence, in this special case, $H = A \otimes F$.)

Proof: By Theorem 8.14, it is sufficient to prove exactness at $A \otimes F$. Let $\{x_k\}$ be a basis of F; then $F = \sum_k^{\oplus} Rx_k$. By Theorem 8.5, $A \otimes F = \sum_k^{\oplus} A \otimes Rx_k$. Hence, $\alpha \in A \otimes F$ has a representation $a = \sum a_k \otimes x_k$. From $(i \otimes 1)a = 0$ it follows that $\sum i(a_k) \otimes x_k = 0$.

By Theorem 8.7, the isomorphism $iA \otimes Rx_k = iA$ is given by $i(a_k) \otimes rx_k = ri(a)$. Hence, for all k, $i(a_k) = 0$; and then $a_k = 0$ since i is a monomorphism. It follows that $a = 0$. Hence, $i \otimes 1$ is a monomorphism. By Theorem 8.5,

$$B \otimes F = \textstyle\sum_k^{\oplus}(B \otimes Rx_k)$$

and, by the definition of the tensor product of functions,

$$(i \otimes 1)(A \otimes F) = \textstyle\sum_k^{\oplus}(iA \otimes Rx_k);$$

from which it follows by use of Theorem 8.7 that $B \otimes Rx_k = B$ and $iA \otimes Rx_k = iA$. Hence, if $i(A)$ is a direct summand of B, then $(i \otimes 1)(A \otimes F)$ is a direct summand of $B \otimes F$.

9

The Functor Hom

It was proved in Chapter 1 that the R-homomorphisms of an R-module A into an R-module B form an R-module $\text{Hom}(A, B)$. Let $\phi: A' \longrightarrow A$ and $\psi: B \longrightarrow B'$ be R-homomorphisms. The function which associates with each $f \in \text{Hom}(A, B)$ the homomorphism $\psi f \phi \in \text{Hom}(A', B')$ is to be designated by $\text{Hom}(\phi, \psi)$. Since $\psi(f + g)\phi = \psi f \phi + \psi g \phi$ and $\psi(rf)\phi = r\psi f \phi$ for $r \in R$, the function $\text{Hom}(\phi, \psi)$ is a homomorphism of $\text{Hom}(A, B)$ into $\text{Hom}(A', B')$. Hom is called *contravariant* in the first argument and *covariant* in the second because

$$\text{Hom}(\phi_1, \psi_1)\,\text{Hom}(\phi_2, \psi_2) = \text{Hom}(\phi_2\phi_1, \psi_1\psi_2).$$

Further rules for computation are

$$\operatorname{Hom}(\phi_1 + \phi_2, \psi) = \operatorname{Hom}(\phi_1, \psi) + \operatorname{Hom}(\phi_2, \psi),$$
$$\operatorname{Hom}(\phi, \psi_1 + \psi_2) = \operatorname{Hom}(\phi, \psi_1) + \operatorname{Hom}(\phi, \psi_2),$$

and

$$\operatorname{Hom}(r\phi, \psi) = r\operatorname{Hom}(\phi, \psi) = \operatorname{Hom}(\phi, r\psi) \text{ for } r \in R.$$

Particularly important are the functions $\bar{\phi} = \operatorname{Hom}(\phi, \mathrm{Id})$ in case $B = B'$, and $\tilde{\psi} = \operatorname{Hom}(\mathrm{Id}, \psi)$ in case $A = A'$. For these, $\bar{\phi} f = f\phi$ and $\tilde{\psi}(f) = \psi f$. The rules for computation yield directly

$$\bar{\phi}_1 + \bar{\phi}_2 = \overline{\phi_1 + \phi_2},\ \widetilde{\psi_1 + \psi_2} = \tilde{\psi}_1 + \tilde{\psi}_2,\ \bar{\phi}_1\bar{\phi}_2 = \overline{\phi_2\phi_1},$$
$$\tilde{\psi}_1\tilde{\psi}_2 = \widetilde{\psi_1\psi_2},\ r\bar{\phi}_1 = \overline{r\phi_1}, \text{ and } r\tilde{\psi}_1 = \widetilde{r\psi_1} \text{ for } r \in R.$$

9.1. THEOREM: *Let* $A \xrightarrow{i} A' \xrightarrow{j} A'' \longrightarrow 0$ *be exact. Then the sequence*

$$\operatorname{Hom}(A, B) \xleftarrow{\bar{i}} \operatorname{Hom}(A', B) \xleftarrow{\bar{j}} \operatorname{Hom}(A'', B) \longleftarrow 0$$

is exact.

Proof:

(1) If $f \in \operatorname{Ker} \bar{j}$, then $\bar{j}(f) = fj = 0$, and hence, $f = 0$, since j is an epimorphism. Therefore, $\bar{j}$ is a monomorphism.

(2) $\bar{i}\bar{j} = \overline{ji} = \bar{0} = 0$.

(3) Let $f \in \operatorname{Ker} \bar{i}$, that is $\bar{i}(f) = fi = 0$. The problem is to find $g \in \operatorname{Hom}(A'', B)$ for which $\bar{j}(g) = gj = f$ and therefore, $gj(a') = f(a')$ for all $a' \in A'$. To $a'' \in A''$ choose $a' \in A'$ for which $j(a') = a''$ and define $g(a'') = f(a')$. Then g is well defined since a' is determined up to an element $i(a)$ and $fi(a) = 0$. Since $gj(a') = g(a'') = f(a')$, it follows that $gj = f$.

9.2. THEOREM: *Let* $0 \longrightarrow B \xrightarrow{i} B' \xrightarrow{j} B''$ *be exact; then the sequence*

$$0 \longrightarrow \operatorname{Hom}(A, B) \xrightarrow{\tilde{i}} \operatorname{Hom}(A, B') \xrightarrow{\tilde{j}} \operatorname{Hom}(A, B'')$$

is exact.

Proof:

(1) If $f \in \operatorname{Ker} \tilde{i}$, then $\tilde{i}(f) = if = 0$, and hence, $f = 0$ since i is a monomorphism. Therefore, $\tilde{i}$ is a monomorphism.

(2) $\tilde{j}\,\tilde{i} = \widetilde{ji} = \tilde{0} = 0$.

(3) Let $\tilde{f} \in \operatorname{Ker} \tilde{j}$; that is, $\tilde{j}(f) = jf = 0$. The problem is to find $g:A \longrightarrow B$ for which $\tilde{i}(g) = f$, and therefore, $ig = f$. To $a \in A$ there is, since $jf(a) = 0$ and the given sequence is exact at B', a unique $b \in B$ with $ib = f(a)$. Define g by $g(a) = b$. Then $g \in \operatorname{Hom}(A, B)$ and $ig(a) = ib = f(a)$ for all $a \in A$. Therefore, $\widetilde{ig} = f$.

9.3. THEOREM: *If* $A = \sum_i^{\oplus} A_i$, *then* $\operatorname{Hom}(A, B) \cong \prod_i \operatorname{Hom}(A_i, B)$.

Proof: Denote the restriction of $f \in \operatorname{Hom}(A, B)$ to A_i by f_i. To f associate the function $(\cdots, f_i \cdots) \in \prod_i \operatorname{Hom}(A_i, B)$. This is an isomorphism of $\operatorname{Hom}(A, B)$ on $\prod_i \operatorname{Hom}(A_i, B)$. For to each set of functions $\{f_i:A_i \longrightarrow B\}$, $i \in I$, there is an $f:A \longrightarrow B$ which is defined by $f(a) = \sum f_i(a_i)$ for $a = \sum a_i \in A$. Since the sum is direct and only finitely many of the a_i are non-zero, f is well defined.

9.4. THEOREM: *If* $B = \prod_k B_k$, *then* $\operatorname{Hom}(A, B) = \prod_k \operatorname{Hom}(A, B_k)$.

Proof: Let $p_k:B \longrightarrow B_k$ be the projection on the kth component. If $f \in \operatorname{Hom}(A, B)$, let $f_k = p_k f:A \longrightarrow B_k$. To f associate $(\cdots, f_k, \cdots) \in \prod_k \operatorname{Hom}(A, B_k)$. This is an isomorphism of $\operatorname{Hom}(A, B)$ onto $\prod_k \operatorname{Hom}(A, B_k)$; for, to each set of functions $\{f_k:A_k \longrightarrow B\}$, $k \in I$, there is an f which is given by $f(a) = (\cdots, f_k(a), \cdots)$.

9.5. *Let* $0 \longrightarrow A \xrightarrow{i} A' \xrightarrow{j} A'' \longrightarrow 0$ *be exact and let* iA *be a direct summand of* A'. *Then*

$$0 \longleftarrow \operatorname{Hom}(A, B) \xleftarrow{\bar{i}} \operatorname{Hom}(A', B) \xleftarrow{\bar{j}} \operatorname{Hom}(A'', B) \longleftarrow 0$$

is exact and $\bar{j}(\operatorname{Hom}(A'', B))$ *is a direct summand of* $\operatorname{Hom}(A', B)$.

Proof: By Theorem 9.1, it is sufficient for the exactness to prove that $\bar{i}$ is an epimorphism. Let $f \in \operatorname{Hom}(A, B)$. Write $A' = iA \oplus C$ and define $g \in \operatorname{Hom}(A', B)$ by $g(ia + c) = f(a)$. Since the sum is direct and i is a monomorphism, g is well defined. For $a \in A$, it follows that $gi(a) = g(ia) = f(a)$, and hence, that $\bar{i}(g) = gi = f$. Then $\bar{i}$ is an epimorphism. By Theorem 9.3,

$$\operatorname{Hom}(A', B) \cong \operatorname{Hom}(iA \oplus C, B) \cong \operatorname{Hom}(iA, B) \oplus \operatorname{Hom}(C, B).$$

Since $\operatorname{Im} \bar{j} = \operatorname{Ker} \bar{i}$, it follows that $\operatorname{Im} \bar{j} = \operatorname{Hom}(C, B)$.

A similar proof will establish the next theorem.

9.6. THEOREM: *Let* $0 \longrightarrow B \xrightarrow{i} B' \xrightarrow{j} B'' \longrightarrow 0$ *be exact, and let iB be a direct summand of B'. Then*

$$0 \longrightarrow \operatorname{Hom}(A, B) \xrightarrow{\tilde{i}} \operatorname{Hom}(A, B') \xrightarrow{\tilde{j}} \operatorname{Hom}(A, B'') \longrightarrow 0$$

is exact and Hom(A, B) *is a direct summand of* Hom(A, B').

Exercise: Let $A \xrightarrow{i} A' \longrightarrow A'' \longrightarrow 0$ and $0 \longrightarrow B \longrightarrow B' \xrightarrow{q} B''$ be exact. Then there is a diagram analogous to the "nine-diagram" and

$$0 \longrightarrow \operatorname{Hom}(A'', B) \longrightarrow \operatorname{Hom}(A', B') \xrightarrow{\bar{q} \oplus \tilde{i}} \operatorname{Hom}(A', B'') \oplus \operatorname{Hom}(A, B'')$$

is exact.

R,S-Modules

Let R and S be two commutative rings with unit elements. Let M be both an R-module and an S-module.

Definition: M is an *R,S-module* if and only if $r(sm) = s(rm)$ for all $r \in R$, $s \in S$ and $m \in M$.

For instance, each R-module is an R-Z-module and an R, R-module.

Definition: Let M, N be two R, S-modules. A function $f: M \longrightarrow N$ is called an *R, S-homomorphism* if and only if f is both an R-homomorphism and an S-homomorphism.

Until now the ring R has usually been fixed in considerations of $A \otimes B$ and Hom(A, B), and consequently it has usually been suppressed in the notation. Where the ring matters, the notation will indicate the ring involved, as in $A \otimes_R B$ and $\operatorname{Hom}_R(A, B)$.

9.7. THEOREM: *Let A be an R-module and B an R,S-module. Then* $A \otimes_R B$, $\operatorname{Hom}_R(A, B)$ *and* $\operatorname{Hom}_R(B, A)$ *can be interpreted as R,S-modules by defining for* $s \in S$:

(a) $s(\sum a \otimes_R b) = \sum(\alpha \otimes_R sb)$,
(b) $(sf)(a) = s(f(a))$,
(c) $(sg)(b) = g(sb)$,

where $a \in A$, $b \in B$, $f \in \operatorname{Hom}_R(A, B)$ *and* $g \in \operatorname{Hom}_R(B, A)$.

Proof: Only in the case of (a) is there a problem about whether the symbols are well defined: The function $h: A \times B \longrightarrow A \otimes_R B$ given by $h(a, b) = a \otimes_R sb$ is R-bilinear. Hence, there is a homomorphism $\tilde{h}: A \otimes_R B \longrightarrow A \otimes_R B$ for which $\tilde{h}(a \otimes_R b) = a \otimes_R sb$. Thus the symbols of (a) are well defined. That the relevant modules are S-modules is clear. That they are R-S-modules follows from

$$r(s(a \otimes_R b)) = r(a \otimes sb) = ra \otimes sb = s(r(a \otimes b)),$$
$$r(sf))(a) = r(sf(a)) = s(rf(a))) = (s(rf))(a), \text{ and}$$
$$(r(sg))(b) = (sg)(rb) = g(s(rb)) = g(r(sb)) = (rg)(sb) = (s(rg))(b).$$

9.8. THEOREM: *Let $f: A \longrightarrow A'$ be an R-homomorphism of the R-modules A and A' and let $g: B \longrightarrow B'$ be an R,S-homomorphism of the R,S-modules B and B'. Then $f \otimes_R g: A \otimes B \longrightarrow A' \otimes B'$, $\text{Hom}_R(f, g): \text{Hom}_R(A', B) \longrightarrow \text{Hom}_R(A, B')$ and $\text{Hom}_R(g, f)$: $\text{Hom}_R(B', A) \longrightarrow \text{Hom}_R(B, A')$ are R,S-homomorphisms.*

Proof: The functions in question are R-homomorphisms. Therefore, it needs only to be proved that they are S-homomorphisms: Let $s \in S$. Then,

$$\begin{aligned}(f \otimes g)(s(a \otimes b)) &= (f \otimes g)(a \otimes sb) = f(a) \otimes g(sb) \\ &= f(a) \otimes sg(b) = s((f \otimes g)(a \otimes b)).\end{aligned}$$

If $h': A' \longrightarrow B$ and $a \in A$, then

$$\begin{aligned}&(\text{Hom}_R(f, g)(sh'))(a) \\ &\quad = (g(sh')f)(a) = g((sh')(f(a))) = g(s(h'f(a))) \\ &\quad = s(g(h'f(a))) = s(gh'f(a)) = (s\,\text{Hom}_R(f, g))(h')(a).\end{aligned}$$

Finally, if $h: B' \longrightarrow A$ and $b \in B$, then

$$\begin{aligned}(\text{Hom}_R(g, f)(sh))(b) &= (f(sh)g)(b) = fh(sg(b)) = fhg(sb) \\ &= (s(fhg))(b) = (s\,\text{Hom}_R(g, f))(h)(b).\end{aligned}$$

9.9. Generalized Associative Law: *Let A be an R-module, B an R,S-module, and C an S-module. Then there is an R,S-isomorphism of $(A \otimes_R B) \otimes_S C$ on $A \otimes_R (B \otimes_S C)$ under which $(a \otimes_R b) \otimes_S c$ goes into $a \otimes_R (b \otimes_S c)$.*

Proof: For a fixed $c \in C$ the function $f_c: B \longrightarrow B \otimes_S C$, defined by $f_c(b) = b \otimes_S c$, is an R, S-homomorphism. It was proved in 9.8

that $(1 \otimes_R f_c): A \otimes_R B \longrightarrow A \otimes_R (B \otimes_S c)$ is an R,S-homomorphism. By definition

$$(1 \otimes_R f_c)(a \otimes_R b) = a \otimes_R (b \otimes_S c).$$

This defines a function $F:(A \otimes_R B) \otimes C \longrightarrow A \otimes_R (B \otimes_S C)$.

Since F is S-bilinear, there is an S-homomorphism $\tilde{F}:(A \otimes_R B) \otimes_S C \longrightarrow A \otimes_R (B \otimes_S C)$, for which

$$\tilde{F}((a \otimes_R b) \otimes_S c) = a \otimes_R (b \otimes_S c).$$

For $r \in R$,

$$\begin{aligned}\tilde{F}(r((a \otimes_R b) \otimes_S c) &= \tilde{F}((ra \otimes_R b) \otimes_S c) = ra \otimes_R (b \otimes_S c)\\ &= r(a \otimes_R (b \otimes_S c) = r\tilde{F}(a \otimes_R (b \otimes_S c).\end{aligned}$$

Consequently, $\tilde{F}$ is an R,S-homomorphism. An interchange of R and S shows the existence of an R,S-homomorphism $\tilde{G}$ for which

$$\tilde{G}(a \otimes_R (b \otimes_S c)) = (a \otimes_R b) \otimes_S c.$$

It follows that $\tilde{F}\tilde{G} = \text{Id}$ and $\tilde{G}\tilde{F} = \text{Id}$. Therefore, $\tilde{F}$ is an R,S-isomorphism.

There is an analogous theorem for the function Hom which will not be needed. It states: *Let A be an R-module, B an R,S-module, and C an S-module. Then there is an R,S-isomorphism*

$$\tilde{F}: \text{Hom}_R(A, \text{Hom}_S(B, C)) \longrightarrow \text{Hom}_S(B \otimes_R A, C)$$

for which

$$\tilde{F}(f)(b \otimes_R a) = f(a)b$$

for $f: A \longrightarrow \text{Hom}_S(B, C)$, $a \in A$ and $b \in B$.

Quotient Modules

Definition: The *torsion submodule* $T(A)$ of the R-module A is the set of all $a \in A$ for which there is an $r \in R$ such that $r \neq 0$ and $ra = 0$. Clearly, $T(A)$ is a submodule of A. In case $T(A) = \phi$, A is called *torsion free*. The factor module $A/T(A)$ is torsion free but is not in general free. A is called a *torsion module* when $T(A) = A$.

Let R be an integral domain and K its quotient field. Since K is an R,K-module, $A \otimes_R K$ is an R,K-module and therefore a vector space over K. The dimension of the K-vector space $A \otimes_R K$ is called the *rank* of A. Let A be embedded in $A \otimes_R K$ under the R-homomorphism $\phi: A \longrightarrow A \otimes_R K$ defined by $\phi(a) = a \otimes_R 1$. The nature of this embedding will be the object of study in the rest of Chapter 9.

In the case that R is a principal ideal ring and A is a finitely generated R-module, there is a theorem, analogous to the principal theorem for abelian groups, which shows that $A \cong T(A) \oplus F$ where F is a free (but not uniquely determined) module isomorphic to $A/T(A)$, and the number of generators of F is the rank of A. The embedding ϕ has $T(A)$ as its kernel.

Warning! If R is not a principal ideal ring or A is not finitely generated, $T(A)$ does not need to be a direct summand of A. In these cases, $A/T(A)$ does not need to be free.

Counter Example: The Z-module Q has infinitely many generators. Clearly, $T(Q) = 0$. However, Q is not free. Since $Q \otimes_Z Q = Q$, the rank of Q is 1.

Counter Example: Let p be a given prime number and let C_n be the cyclic group of order p^n. *In* $G = \prod_{k=0}^{n} C_k$, $T(G)$ *is not a direct summand.* This can be seen as follows: From $G = T(G) \oplus H$, it follows that $H \cong G/T(G)$. Let a be an element $(\cdots, p^{[p/z]}, \cdots)$ which is not congruent to zero mod $T(G)$. To each natural number r, there is a $b \in G$ such that $a \equiv p^r b$ mod $T(G)$; then there is also an $a' \in H$ with $a' \neq 0$ and to each r there is a b' with $a' = p^r b'$. The only element of G having this property, however, is $a = 0$. This contradiction shows that $T(G)$ is no direct summand of G.

Other counter examples are to be found in Kaplansky, *Infinite Abelian Groups*.

9.10. THEOREM: *Let R be an integral domain, K its quotient field, and A an R-module. Then $T(A)$ is the kernel of the homomorphism $\phi: A \longrightarrow A \otimes_R K$ given by $\phi(a) = a \otimes 1$ for $a \in A$.*

Proof: To $a \in T(A)$ there is $r \in R$ for which $r \neq 0$ and $ra = 0$. Consequently, $\phi(a) = a \otimes 1 = ra \otimes 1/r = 0$, and therefore, $a \in \operatorname{Ker} \phi$. Conversely, let $\phi(a) = a \otimes 1 = 0$. It was proved in 8.11 that A is a factor module of a free module F, say $A = F/L$. By 8.5,

$$F \otimes_R K = \sum^{\oplus} Rx_k \otimes_R K,$$

where the x_k are the elements of a basis of F. Now let $f = \sum r_k x_k$, where $f \otimes 1 = 0$. Then $\sum r_k x_k \otimes 1 = 0$ and, for each k, $r_k x_k \otimes 1 = 0$ since the sum is direct. By 8.7, there is an isomorphism of $Rx_k \otimes K$ on K which maps $r_k x_k \otimes a_k$ on ra_k. Hence, $r_k = 0$ for each k, and $f = 0$. By Theorem 8.9, the tensor product $A \otimes_R K = (F/L) \otimes_R K$ can be taken to be $(F \otimes K)/H$, where H is generated by all $l \otimes \alpha$, $l \in L$, $\alpha \in K$. Let $h = \sum_i l_i \otimes \alpha_i$, where $l_i \in L$ and $\alpha_i \in K$. If $\beta \in R$ is the common denominator of the a_i, then $h = l \otimes (1/\beta)$, where $l = \sum \beta\alpha_i l_i \in L$. Hence the elements $h \in H$ have the form $h = l \otimes (1/\beta)$, where $l \in L$ and $\beta \in R$, $\beta \neq 0$. Now let $a \in R$ satisfy $\phi(a) = a \otimes 1 = 0$. Then $a = \sum r_k x_k + L$ for certain $r_k \in R$. By 8.10,

$$a \otimes 1 = (\textstyle\sum r_k x_k \otimes 1) + H.$$

By $\phi(a) = 0$ there exists an $h \in H$ for which

$$h = l \otimes 1/\beta = \textstyle\sum r_k x_k \otimes 1.$$

Consequently,

$$(l - \beta \textstyle\sum r_k x_k) \otimes 1 = 0,$$

and therefore,

$$l - \beta \textstyle\sum r_k x_k = 0$$

as above. This means that $\beta\alpha = 0$, from which it follows that $\alpha \in T(A)$.

Remarks:

(1) That $a \in T(A)$ implies $a \otimes \alpha = 0$ for $\alpha \in K$ can be seen as follows: Pick $r \in R$ such that $r \neq 0$ and $ra = 0$. Then

$$a \otimes \alpha = a \otimes (r\alpha/r) = ra \otimes (\alpha/r) = 0.$$

(2) That ($a \otimes \alpha = 0$ for a non-zero α in K) implies ($a \in T(A)$) can be seen as follows: Let $\alpha = r/s$ with $r, s \in R$. Then $s(a \otimes \alpha) = (ra) \otimes 1 = 0$. Consequently, $r\alpha \in T(A)$ and then $a \in T(A)$.

9.11. Corollary:

(1) *The image $\phi(A)$ of A in $A \otimes_R K$ is isomorphic to $A/T(A)$.*

(2) *A is torsion free if and only if ϕ is a monomorphism.*
(3) $A \otimes_R K = (A/T(A)) \otimes_R K$.

Proof: Parts (1) and (2) are obvious. For (3), Theorem 8.9 yields

$$(A/T(A)) \otimes_R K \cong (A \otimes K)/H,$$

where H is generated by all $a \otimes \beta$ for which $a \in T(A)$ and $\beta \in K$. Remark (1) shows that such $a \otimes \beta$ are all zero, and therefore that $H = 0$.

Description of $A \otimes_R K$: Each finite set of elements of K can be written with a common denominator. Hence, each element of $A \otimes K$ has the form $a \otimes (1/r)$, where $a \in A$, $r \in R$, and $r \neq 0$. Furthermore, any two elements can be written with the same r. The equality $a \otimes (1/r) = b \otimes (1/s)$ is equivalent to $\phi(sa - rb) = 0$, which is equivalent to $sa - rb \in T(A)$, and this in turn is equivalent to the statement: There is a $t \in R$ such that $t \neq 0$ and $t(sa - rb) = 0$. It is therefore possible to define the quotients a/r by $a/r = a \otimes (1/r)$. The equation $(a/r) = (b/s)$ then holds if and only if there is a $t \in R$ such that $t \neq 0$ and $t(sa - rb) = 0$. Furthermore,

$$(a/r) + (b/s) = (sa + rb)/rs$$

and

$$\alpha(a/r) = (sa)/(tr)$$

for $\alpha = (s/t) \in K$. The product $A \otimes_R K$ is also called the *quotient module* of A.

9.12. *The operation (functor) $\otimes_R K$ preserves exactness.*

Proof: Let $A \xrightarrow{i} B \xrightarrow{j} C$ be exact. It must be proved that $A \otimes_R K \xrightarrow{i \otimes 1} B \otimes_R K \xrightarrow{j \otimes 1} C \otimes_R K$ is exact.

(1) $(j \otimes 1)(i \otimes 1) = (ji) \otimes 1 = 0$ since $ji = 0$.
(2) The elements of $B \otimes_R K$ have the form $b \otimes (1/r)$, where $b \in B$, $r \in R$, and $r \neq 0$. Let $(j \otimes 1)(b \otimes 1/r) = 0$. Then $j(b) \otimes 1 = 0$ and hence, $j(b) \in T(C)$ by Remark (2) preceding 9.11. Therefore, there is an $s \in R$ such that $s \neq 0$ and $sj(b) = j(sb) = 0$. Hence, there is an $a \in A$ with $i(a) = sb$. It follows that

$$(i \otimes 1)(a \otimes (1/(sr))) = i(a) \otimes (1/(sr)) = b \otimes (1/r).$$

10

Categories and Functors

Definition: Let there be given a class (*class*, not set) of objects, which will be denoted by $A, B, C, \cdots, X, Y, \cdots$. To each ordered pair A, B of the objects let a (possibly empty) set $\mathscr{A}$ be assigned. The elements of $\mathscr{A}(A, B)$ are called *morphisms*, or arrows, and are denoted by $\phi, \psi, \chi, \cdots$. The statement $\phi \in \mathscr{A}(A, B)$ is often written $\phi : A \longrightarrow B$ or $A \xrightarrow{\phi} B$. Furthermore, let there be given to each three objects A, B, C, a composition $\mathscr{A}(BC) \cdot \mathscr{A}(A, B) \subset \mathscr{A}(A, C)$. For $\phi : A \longrightarrow B$ and $\psi : B \longrightarrow C$, this composition is written $\psi\phi : A \longrightarrow C$. The given class of objects together with the corresponding morphisms is called a *category* provided that the following conditions (axioms for a category) hold:

(K1). If $A \xrightarrow{\phi} B \xrightarrow{\psi} C \xrightarrow{\chi} D$ then $(\chi\psi)\phi = \chi(\psi\phi)$.

(K2). Let A be an object. Then there exists $e_A \in \mathscr{A}(A, A)$ such that $\phi e_A = \phi = e_B\phi$ for each $\phi \in \mathscr{A}(A, B)$.

The element e_A is called an *identity*. If e'_A is a second identity, then $e'_A = e_A e'_A = e_A$; and e_A is thus uniquely determined.

The arrow $\phi: A \longrightarrow B$ is called an *equivalence* if and only if there is a $\psi: B \longrightarrow A$ such that $\psi\phi = e_A$ and $\phi\psi = e_B$. If $\psi'\phi = e_A$ and $\phi\psi' = e_B$ hold for $\psi': B \longrightarrow A$ then

$$\psi = e_A\psi = \psi'\phi = \psi' e_B = \psi'.$$

Hence ψ is uniquely determined by ϕ, if it exists. Then morphism ψ, if if exists, is called the *inverse* of ϕ and is written $\psi = \phi^{-1}$. If ϕ is an equivalence with ϕ^{-1} as inverse, then ϕ^{-1} is also an equivalence and has $(\phi^{-1})^{-1} = \phi$ as inverse. If $\phi: A \longrightarrow B$ and $\psi: B \longrightarrow C$ are equivalences, then $\psi\phi$ is also an equivalence and $(\psi\phi)^{-1} = \phi^{-1}\psi^{-1}$, since

$$\psi\phi\phi^{-1}\psi^{-1} = \psi e_B \psi^{-1} = \psi\psi^{-1} = e_C$$

and

$$\phi^{-1}\psi^{-1}\psi\phi = \phi^{-1}e_B\phi = e_A.$$

A category $\mathscr{K}$ is called *linear* if and only if each $\mathscr{A}(A, B)$ is an additive group and the distributive laws

$$(\psi_1 + \psi_2)\phi = \psi_1\phi + \psi_2\phi \quad \text{and} \quad \psi(\phi_1 + \phi_2) = \psi\phi_1 + \psi\phi_2$$

hold for $\phi, \phi_1, \phi_2: A \longrightarrow B$ and $\psi, \psi_1, \psi_2: B \longrightarrow C$. In a linear category, the sets $\mathscr{A}(A, B)$ are not empty since $0 \in \mathscr{A}(A, B)$.

The space pairs and their continuous functions form a nonlinear category. Examples of linear categories are:

(1) The class of objects consists of the R-modules; the morphisms are their homomorphisms.

(2) The class of objects consists of the R-chain complexes of R-modules A_q together with those of their R-homomorphisms ∂_q that satisfy $\partial_{q+1}\partial_q = 0$, where

$$\cdots \longrightarrow A_{q+1} \xrightarrow{\partial_{q+1}} A_q \xrightarrow{\partial_q} A_{q-1} \longrightarrow \cdots.$$

Let A, B be chain complexes. Then $\mathscr{A}(A, B)$ consists of all chain homomorphisms $f^{\#}: A \longrightarrow B$. Such a chain homomorphism is not a function, but a set of homomorphisms

$$\begin{array}{ccccccccc} \cdots \longrightarrow & A_{q+1} & \xrightarrow{\partial_{q+1}} & A_q & \xrightarrow{\partial_q} & A_{q-1} & \longrightarrow \cdots \\ & \downarrow{\scriptstyle f^\#_{q+1}} & & \downarrow{\scriptstyle f^\#_q} & & \downarrow{\scriptstyle f^\#_{q-1}} & \\ \cdots \longrightarrow & B_{q+1} & \xrightarrow{\partial_{q+1}} & B_q & \xrightarrow{\partial_q} & B_{q-1} & \longrightarrow \cdots, \end{array}$$

where the $f^\#_q$ are R-homomorphisms and the rectangles commute.

(3) Each R-chain complex A has a corresponding set $H(A)$ of homology groups $H_q(A), q \in Z$. These sets $H(A)$ are the objects. Each chain homomorphism $f^\#:A \longrightarrow B$ corresponds to a set f_* of module homomorphisms $f_{*q}:H_q(A) \longrightarrow H_q(B)$. The set $\mathscr{A}(H(A), H(B))$ consists of all f_*.

Definition: Let $\mathscr{K}$ and $\mathscr{K}'$ be two categories. A *covariant functor* T (of one variable) of $\mathscr{K}$ in $\mathscr{K}'$ is a function which associates with each object $A \in \mathscr{K}$ an object $B \in \mathscr{K}'$ and with each morphism $\phi:A \longrightarrow B$ a morphism $T(\phi):T(A) \longrightarrow T(B)$ such that $T(e_A) = e_{T(A)}$ and $T(\psi\phi) = T(\psi)T(\phi)$.

A *contravariant functor* T of $\mathscr{K}$ in $\mathscr{K}'$ is a function which associates with each object $A \in \mathscr{K}$ an object $T(A) \in \mathscr{K}'$ and with each morphism $\phi:A \longrightarrow B$ a morphism $T(\phi):T(B) \longrightarrow T(A)$ such that $T(e_A) = e_{T(A)}$ and $T(\psi\phi) = T(\phi)T(\psi)$.

If $\mathscr{K}$ and $\mathscr{K}'$ are linear categories, then a functor T of $\mathscr{K}$ in $\mathscr{K}'$ is called *additive* if and only if

$$T(\phi + \psi) = T(\phi) + T(\psi)$$

for $\phi, \psi \in \mathscr{A}(A, B)$.

Application of a functor T to a commutative diagram

$$\begin{array}{ccc} A & \xrightarrow{\phi} & B \\ \downarrow{\scriptstyle \alpha} & & \downarrow{\scriptstyle \beta} \\ C & \xrightarrow{\psi} & D \end{array}$$

yields another commutative diagram. In case T is covariant, the image diagram is

$$\begin{array}{ccc} T(A) & \xrightarrow{T(\phi)} & T(B) \\ \downarrow{\scriptstyle T(\alpha)} & & \downarrow{\scriptstyle T(\beta)} \\ C & \xrightarrow{T(\psi)} & D. \end{array}$$

Since

$$T(\beta)T(\phi) = T(\beta\phi) = T(\psi\alpha) = T(\psi)T(\alpha),$$

the diagram commutes. In case T is contravariant, the image diagram is

$$\begin{array}{ccc} T(A) & \xleftarrow{T(\phi)} & T(B) \\ \uparrow{\scriptstyle T(\alpha)} & & \uparrow{\scriptstyle T(\beta)} \\ T(C) & \xleftarrow{T(\psi)} & T(D). \end{array}$$

Since

$$T(\phi)T(\beta) = T(\beta\phi) = T(\psi\alpha) = T(\alpha)T(\psi),$$

this diagram also commutes.

Example: Let $\mathscr{K}$ be the linear category of Example (2) and let $\mathscr{K}'$ be that of Example (3). The mapping T which associates with each chain complex the set of its homology groups, and with each $f^{\#}$ its image $T(f^{\#}) = f_*$ is a covariant, additive functor of $\mathscr{K}$ in $\mathscr{K}'$: For by Lemma 1.2,

$$T(Id) = Id,\ T(g^{\#}f^{\#}) = (g^{\#}f^{\#})_* = g_*f_* = T(g^{\#})T(f^{\#}),$$

and

$$T(f^{\#} + g^{\#}) = T(f^{\#}) + T(g^{\#}).$$

Definition: A sequence $\cdots \longleftarrow A^{q+1} \xleftarrow{\delta^{q+1}} A^q \xleftarrow{\delta^q} A^{q-1} \longleftarrow \cdots$ of R-modules A^q, $q \in Z$, and R-homomorphisms δ^q such that $\delta^q\delta^{q+1} = 0$, is called an *R-cochain complex*. By means of a renumbering, each cochain complex can be interpreted as a chain complex. Therefore, the same theorems hold for chain and cochain complexes (this does not say that homology and cohomology groups are the same).

10.1. THEOREM: *Let R and S be two rings, and let T be an additive covariant (contravariant) functor of the category of S-modules into the category of R-modules. Then T induces in a natural way a functor T' of the category of S-chain complexes into the category of R-chain complexes.*

Proof: Let

$$\cdots \longrightarrow A_{q+1} \xrightarrow{\partial_{q+1}} A_q \xrightarrow{\partial_q} A_{q-1} \longrightarrow \cdots$$

be an S-chain complex A. With each $\partial_q : A_q \longrightarrow A_{q-1}$ is associated a $T(\partial_q)$. When T is covariant, $T(\partial_q):T(A_q) \longrightarrow T(A_{q-1})$; when T is contravariant, $T(\partial_q):T(A_{q-1}) \longrightarrow T(A_q)$. Since T is additive, $T(\partial_{q-1}\partial_q) = 0$. When T is covariant,

$$T(\partial_{q-1})T(\partial_q) = T(\partial_{q-1}\partial_q) = 0.$$

Hence,

$$\cdots \longrightarrow T(A_{q+1}) \xrightarrow{T(\partial_{q+1})} T(A_q) \xrightarrow{T(\partial_q)} T(A_{q-1}) \longrightarrow \cdots$$

is an R-chain complex $T'(A)$. When T is contravariant,

$$T(\partial_q)T(\partial_{q-1}) = T(\partial_{q-1}\partial_q) = 0$$

and, therefore,

$$\cdots \longleftarrow T(A_{q+1}) \xleftarrow{T(\partial_{q+1})} T(A_q) \xleftarrow{T(\partial_q)} T(A_{q-1}) \longleftarrow \cdots$$

is a cochain complex $T'(A)$. The elements of $\mathscr{A}(A, B)$ are the S-chain homomorphisms $f^{\#}:A \longrightarrow B$. The image $T'(f^{\#})$ is defined by $(T'(f^{\#}))_q = T(f^{\#}_q)$. Clearly the $(T'(f^{\#}))_q$ are R-homomorphisms, and they yield a chain homomorphism since the functor T carries commutative diagrams into commutative diagrams.

Two chain homomorphisms $f^{\#}:A \longrightarrow B$ and $g^{\#}:A \longrightarrow B$ were called chain homotopic if there existed homomorphisms $D_q:A_q \longrightarrow B_{q+1}$ such that

$$f^{\#}_q - g^{\#}_q = \partial_{q+1}D_q + D_{q-1}\partial_q.$$

Application of T to this equation yields

$$\begin{aligned}(T'(f^{\#}))_q - (T'(g^{\#}))_q &= T(f^{\#}_q) - T(g^{\#}_q)\\ &= T(\partial_{q+1})T(D_q) + T(D_{q-1})T(\partial_q).\end{aligned}$$

Thus, if $f^{\#}$ and $g^{\#}$ are chain homotopic, then $T'(f^{\#})$ and $T'(g^{\#})$ are also chain homotopic. This has proved the next theorem.

10.2. THEOREM: *If T is an additive functor of the category of*

S-modules in the category of R-modules, then the induced functor T' preserves each chain homotopy (and each cochain homotopy).

Exactness is not generally preserved by a functor. However, Theorems 9.5 and 9.6 can be generalized.

Definition: Let A, A_α, and B_α be R-modules and let α be an element of some index set. A set of homomorphisms $i_\alpha:A_\alpha \longrightarrow A$ is called an *injective representation of A as a direct sum* $\sum^{\oplus} i_\alpha(A_\alpha)$ if and only if each element $a \in A$ has a unique representation $a = \sum_\alpha i_\alpha(a_\alpha)$, where the sum contains only finitely many non-zero terms.

Definition: A set of homomorphisms $p_\alpha:A \longrightarrow B_\alpha$ is called a *projective representation of A as a direct product of the* B_α if and only if to each $b_\alpha \in B_\alpha$ there is a unique $a \in A$ such that $p_\alpha(a) = b_\alpha$.

Remark: For finite index sets, it is true that a direct sum is a direct product; however, an injective representation is not a projective representation. The first concept is concerned with subgroups, the second with factor groups.

10.3. Lemma: *Let A, A_α and B_α be R-modules. Let $i_\alpha:A_\alpha \longrightarrow A$ and $p_\alpha:A \longrightarrow B_\alpha$ for $\alpha = 1, 2, \cdots, n$. Let $p_\alpha i_\alpha = k_\alpha$, $k_\alpha:A_\alpha \longrightarrow B_\alpha$ and let k_α be an isomorphism. Further, let $p_\alpha i_\beta = 0$ for $\alpha \neq \beta$ and let* $\sum_\alpha i_\alpha k_\alpha^{-1} p_\alpha = \text{Id}$. *Then the i_α define an injective representation of A as a direct sum, and the p_α define a projective representation of A as a direct product.*

Proof: By hypothesis, $a = \text{Id}\ a = \sum_\alpha i_\alpha k_\alpha^{-1} p_\alpha(a)$ for all $a \in A$. This means that each a has a representation $a = \sum_\alpha i_\alpha(a_\alpha)$, where $a_\alpha \in A_\alpha$. It follows from the hypothesis that $p_\beta \alpha = p_\beta i(a_\beta) = k_\beta(a_\beta)$. Since k_β is an isomorphism, the a_β are uniquely determined by a. In particular, each i_α is a monomorphism.

Now suppose that $b_\alpha \in B_\alpha$ has been preassigned. Let $a = \sum_\alpha i_\alpha k_\alpha^{-1}(b_\alpha)$. Then $p_\beta(a) = b_\beta$. Hence, p_β is an epimorphism. If $p_\alpha(a) = 0$ for all α, then

$$a = \text{Id}\ a = \textstyle\sum_\alpha i_\alpha k_\alpha^{-1} p_\alpha(a) = 0.$$

Therefore, there is to the prescribed b_α at most one $a \in A$ such that $p_\alpha(a) = b_\alpha$.

4

10.4. Lemma: *Let* $n = 2$. *In the diagram*

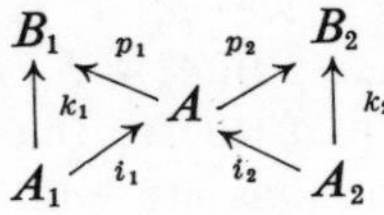

let $k_1 = p_1 i_1$ *and* $k_2 = p_2 i_2$ *be isomorphisms. Suppose that* $p_1 i_2 = 0$ *and* $p_2 i_1 = 0$. *Then*

$$i_1 k_1^{-1} p_1 + i_2 k_2^{-1} p_2 = \mathrm{Id}$$

if and only if at least one of the diagonals is exact.

Proof:

(1) Let $i_1 k_1^{-1} p_1 + i_2 k_2^{-1} p_2 = \mathrm{Id}$. By hypothesis, $p_2 i_1 = 0$. The kernel of p_2 remains to be investigated. From $p_2(a) = 0$, it follows that

$$a = \mathrm{Id}\; a = i_1 k_1^{-1} p_1(a),$$

and then that $a \in i_1 A_1$. Therefore, the sequence $A_1 \xrightarrow{i_1} A \xrightarrow{p_2} B_2$ is exact.

(2) Let $A_1 \xrightarrow{i_1} A \xrightarrow{p_2} B_2$ be exact. Let $a \in A$, where $p_1(a) = 0$ and $p_2(a) = 0$. There is an $a_1 \in A_1$ such that $a = i_1(a_1)$. From $0 = p_1 i_1(a_1) = k_1(a_1)$ it follows that $a_1 = 0$. Therefore, $a = 0$. For $\alpha = 1, 2$ it is clear that

$$p_\alpha(\mathrm{Id} - i_1 k_1^{-1} p_1 - i_2 k_2^{-1} p_2)(x) = p_\alpha(x) - p_\alpha(x) = 0,$$

where $x \in A$. Therefore, $\mathrm{Id} = i_1 k_1^{-1} p_1 + i_2 k_2^{-1} p_2$.

10.5. Lemma: *The sequence* $0 \longrightarrow A \xrightarrow{i} B \xrightarrow{p} C \longrightarrow 0$ *is exact, and* $i(A)$ *is a direct summand of* B *if and only if there exist homomorphisms* i' *and* p' *such that the diagram*

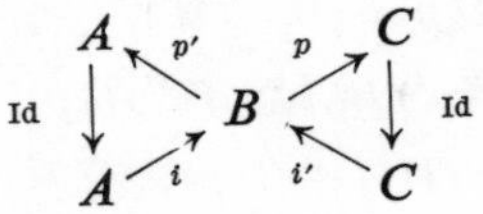

commutes and the equations $p'i' = 0$, $pi = 0$, *and* $ip' + i'p = \mathrm{Id}$ *hold.*

Proof:

(1) If such i' and p' exist, then $A \xrightarrow{i} B \xrightarrow{p} C$ is exact by Lemma 10.4. By Lemma 10.3, i is a monomorphism, $i(A)$ is a direct summand of B, and p is an epimorphism.

(2) Let $0 \longrightarrow A \xrightarrow{i} B \xrightarrow{p} C \longrightarrow 0$ be exact, and let $B = i(A) \oplus B'$. Since $i(A) = \text{Ker } p$, B' is mapped isomorphically on C by p. Now define $p'(i(A) + b') = a$ and $i'(c) = b'$ when $p(b') = c$. Then it is easy to see that the triangles are commutative and the equations are fulfilled.

10.6. THEOREM: *Let A, B, and C be S-modules. Let* $0 \longrightarrow A \xrightarrow{i} B \xrightarrow{p} C \longrightarrow 0$ *be exact and* $i(A)$ *be a direct summand of B. Let T be an additive functor of the S-modules into the R-modules.*

If T is covariant, then $0 \longrightarrow T(A) \xrightarrow{T(i)} T(B) \xrightarrow{T(p)} T(C) \longrightarrow 0$ *is exact and* $T(i)T(A)$ *is a direct summand of* $T(B)$. *If T is contravariant, then* $0 \longleftarrow T(A) \xleftarrow{T(i)} T(B) \xleftarrow{T(p)} T(C) \longleftarrow 0$ *is exact and* $T(p)T(C)$ *is a direct summand of* $T(B)$.

Proof: To i and p, Lemma 10.5 provides i' and p' such that $pi = 0$, $p'i' = 0$, $pi' = \text{Id}$, $p'i = \text{Id}$, and $ip' + i'p = \text{Id}$. If T is covariant, it follows that

$$T(p)T(i) = 0, \ T(p')T(i') = 0,$$
$$T(p')T(i) = \text{Id}, \ T(i)T(p') = \text{Id},$$
$$T(i)T(p') + T(i')T(p) = \text{Id}.$$

Then Lemma 10.5 shows that

$$0 \longrightarrow T(A) \xrightarrow{T(i)} T(B) \xrightarrow{T(p)} T(C) \longrightarrow 0$$

is exact, and that $T(i)T(A)$ is a direct summand of $T(B)$. If T is contravariant, it follows that

$$T(i)T(p) = 0, \ \ T(i')T(p') = 0,$$
$$T(i)T(p') = \text{Id}, \ \ T(i')T(p) = \text{Id},$$
$$T(p')T(i) + T(p)T(i') = \text{Id}.$$

Lemma 10.5 then shows that

$$0 \longleftarrow T(A) \xleftarrow{T(i)} T(B) \xleftarrow{T(p)} T(C) \longleftarrow 0$$

is exact, and that $T(p)T(C)$ is a direct summand of $T(B)$.

11

Categories, Functors, and the Singular Theory

The space pairs (X, A) and the continuous functions $f:(X, A) \longrightarrow (Y, B)$ form a nonlinear category $\mathscr{K}_1$. Suppose that S is a chosen unitary ring and that $\mathscr{K}_2$ is the category of the S-chain complexes. In Chapter 4, an S-chain complex was associated with each space pair (X, A) by the sequence

$$\cdots \longrightarrow S_{q+1}(X, A) \xrightarrow{\partial_{q+1}} S_q(X, A) \xrightarrow{\partial_q} S_{q-1}(X, A) \longrightarrow \cdots.$$

Each continuous function $f:(X, A) \longrightarrow (Y, B)$ induced a chain homomorphism $f^{\#}:S(X, A) \longrightarrow S(Y, B)$. This association is a covariant functor of $\mathscr{K}_1$ into $\mathscr{K}_2$.

Let T be a covariant (contravariant) additive functor of the category of S-modules in the category of R-modules. It was seen in 10.1 that T induces an additive functor T' of $\mathscr{K}_2$ into the category $\mathscr{K}_3$ of R-chain complexes (R-cochain complexes). Let $\mathscr{K}_4$ be the category of Example (3) from the beginning of Chapter 10. In the example which precedes 10.1, there is defined a functor from $\mathscr{K}_3$ into $\mathscr{K}_4$ which associates with the chain complex A its homology group $H(A)$. Composition of these functors yields a functor from $\mathscr{K}_1$ into $\mathscr{K}_4$.

Homology: Let T be covariant. Then the functor of $\mathscr{K}_1$ into $\mathscr{K}_4$ defined by means of T associates with the space pair (X, A) the set $H(T'(S(X, A)))$, which will be abbreviated as $H(X, A)$. The elements of $H(X, A)$ are the homology groups $\operatorname{Ker} T(\partial_q)/\operatorname{Im} T(\partial_{q+1})$ and are denoted by $H_q(X, A)$. The functor of $\mathscr{K}_1$ in $\mathscr{K}_4$ associates with the continuous function $f:(X, A) \longrightarrow (Y, B)$ the set $\{T'(f^{\#})\}_*$, which is abbreviated f_*. The elements of f_* are the R-homomorphisms

$$f_{*q} = (T'(f^{\#}))_{*q}:H_q(X, A) \longrightarrow H_q(Y, B).$$

For covariant functors T, the induced functor of $\mathscr{K}_1$ in $\mathscr{K}_4$ is covariant.

Cohomology: Let T be contravariant. Then T' is contravariant, and so also is the induced functor of $\mathscr{K}_1$ in $\mathscr{K}_4$ obtained by composition. The induced functor associates with the space pair (X, A) the set $H(T'(s(X, A)))$, which again is abbreviated $H(X, A)$. The elements of $H(X,A)$ are the cohomology groups $\operatorname{Ker} T(\partial_{q+1})/\operatorname{Im} T(\partial_q)$, and are denoted by $H^q(X, A)$. The functor of $\mathscr{K}_1$ in $\mathscr{K}_4$ associates with the continuous function $f:(X, A) \longrightarrow (Y, B)$ the set $(T'(f^{\#}))_*$ which is abbreviated f^*. The elements of f^* are the R-homomorphisms

$$(T'(f^{\#}))_{*q} = f^{*q}:H^q(Y, B) \longrightarrow H^q(X, A).$$

The indices have been written as superscripts to indicate that the functor of $\mathscr{K}_1$ in $\mathscr{K}_4$ is contravariant.

Homotopy Theorem: Let $\lambda,\mu:(X, A) \longrightarrow (X \times I, A \times I)$ be the familiar functions $\lambda(x) = (x, 0)$, $\mu(x) = (x, 1)$. Formula 5.13 shows that the induced $\lambda^{\#}$ and $\mu^{\#}$ are chain homotopic. By Theorem 10.2, $T'(\lambda^{\#})$ and $T'(\mu^{\#})$ are also chain homotopic. If T is covariant,

Theorem 5.4 shows that $\lambda_{*q} = \mu_{*q}$ where $\lambda_{*q}, \mu_{*q}:H_q(X, A) \longrightarrow H_q(X \times I, A \times I)$. If T is contravariant, Theorem 5.4 shows that $\lambda^{*q} = \mu^{*q}$, where $\lambda^{*q}, \mu^{*q}:H^q(X \times I, A \times I) \longrightarrow H^q(X, A)$. The considerations preceding Theorem 5.4 now show:

11.0. THEOREM: *If $f,g:(X, A) \longrightarrow (Y, B)$ are homotopic, continuous functions, then $f_* = g_*$ and $f^* = g^*$.*

Excision Theorem: Let $U \subset A$ and $\bar{U} \subset \mathring{A}$. Let $i(X - U, A - U) \longrightarrow (X, A)$ be the injection. The proof of the excision theorem (6.17) contains the construction of a chain homomorphism ϕ such that $\phi i^\# = \text{Id}$, and $i^\# \phi$ is chain homotopic to the identity. If use is made of T', these properties are preserved (the order of the factors is reversed when T is contravariant). Theorem 5.4 now implies that $i_*:H_q(X - U, A - U) \longrightarrow H_q(X, A)$ and $i^*:H^q(X, A) \longrightarrow H^q(X - U, A - U)$ are isomorphisms (in the non-augmented case).

Space Triples: Let (X, A, B) be a space triple, and let

$$0 \longrightarrow S_q(A, B) \xrightarrow{i^\#} S_q(X, B) \xrightarrow{j^\#} S_q(X, A) \longrightarrow 0$$

be the corresponding exact sequence of S-modules of singular chains. Here, $i:(A, B) \longrightarrow (X, B)$ and $j:(X, B) \longrightarrow (X, A)$ are the functions induced by $\text{Id}:X \longrightarrow X$. It will first be proved that $\text{Im}\, i^\#$ is a direct summand of $S_q(X, B) = S_q(X)/S_q(B)$. For $q \geq 0$, $S_q(X)$ was the free S-module generated by the q-simplices σ for which $|\sigma| \subset X$; and $S_q(B)$ was generated by the q-simplices σ for which $|\sigma| \subset B$. Hence, $S_q(X, B)$ is the free module generated by all σ for which $|\sigma| \subset X$, $|\sigma| \not\subset B$, and, analogously, $S_q(A, B)$ is the free S-module generated by all σ for which $|\sigma| \subset A$, and $|\sigma| \not\subset B$. Consequently, $i^\# S_q(A, B)$ is a direct summand of $S_q(X, B)$. If T is covariant, it yields the exact sequence

$$0 \longrightarrow T(S_q(A, B)) \xrightarrow{T(i^\#)} T(S_q(X, B)) \xrightarrow{T(j^\#)} T(S_q(X, A) \longrightarrow 0.$$

If T is contravariant, the same sequence is obtained except that the arrows are in the opposite direction. In the case of homology, Theorem 1.3 yields the exact sequence

$$\cdots \longrightarrow H_q(A, B) \xrightarrow{i_*} H_q(X, B) \xrightarrow{j_*} H_q(X, A) \xrightarrow{\partial_*} H_{q-1}(A, B) \longrightarrow \cdots.$$

In the case of cohomology it yields the exact sequence

$$\cdots H^q(A, B) \xleftarrow{i^*} H^q(X, B) \xleftarrow{j^*} H^q(X, A) \xleftarrow{\delta^*} H^{q-1}(A, B) \cdots,$$

where δ is the usual designation for the coboundary. It should be noted that the $S_q(\cdot,\cdot)$ are S-modules while the $H_q(\cdot,\cdot)$ and $H^q(\cdot,\cdot)$ are R-modules.

Let $f:(X, A, B) \longrightarrow (X', A', B')$ be continuous. Each rectangle of the diagram

$$\begin{array}{ccccccccc} 0 & \longrightarrow & S_q(A, B) & \xrightarrow{i^\#} & S_q(X, B) & \xrightarrow{j^\#} & S_q(X, A) & \longrightarrow & 0 \\ \downarrow & & \downarrow g^\# & & \downarrow h^\# & & \downarrow k^\# & & \downarrow \\ 0 & \longrightarrow & S_q(A', B') & \xrightarrow{i'^\#} & S_q(X', B') & \xrightarrow{j'^\#} & S_q(X', A') & \longrightarrow & 0 \end{array}$$

is commutative when $g^\#$, $h^\#$, and $k^\#$ are the functions induced by f. The commutativity is preserved under an application of T. Now Theorem 1.4 applies to yield, when T is covariant,

$$\begin{array}{ccccccccc} \cdots \longrightarrow & H_q(A,B) & \xrightarrow{i_*} & H_q(X,B) & \xrightarrow{j_*} & H_q(X,A) & \xrightarrow{\partial_*} & H_{q-1}(A,B) & \longrightarrow \cdots \\ & \downarrow & & \downarrow & & \downarrow & & \downarrow & \\ \cdots \longrightarrow & H_q(A',B') & \xrightarrow{i'_*} & H_q(X',B') & \xrightarrow{j'_*} & H_q(X',A') & \xrightarrow{\partial'_*} & H_{q-1}(A',B') & \longrightarrow \cdots \end{array}$$

in which all rectangles are commutative. A contravariant functor T yields an analogous diagram in which the arrows run from right to left, and the stars on the functions are written up rather than down.

This completes the proof of the fundamental theorems for the homology and cohomology groups of the singular theory defined in terms of T.

The Functors $\otimes\, M$ and Hom($, M$) in the Singular Theory

Let S be chosen to be the ring Z of integers. Then the modules $S_q(X, A)$, $q \geq 0$, of the singular theory are free abelian groups generated by the q-simplices σ for which $|\sigma| \subset X$, $|\sigma| \not\subset A$. Let R be a preassigned ring with unit, and let M be a preassigned R-module.

Homology

For the functor T choose $\otimes_Z M$, that is, if C is a Z-module, then $T(C) = C \otimes_Z M$. For a Z-homomorphsim $f:C \longrightarrow D$, it follows that

$$T(f) = f \otimes_Z 1 : C \otimes_Z M \longrightarrow D \otimes_Z M.$$

Since M is an R, Z-module, then $T(C)$ is an R-module and $T(f)$ an R-homomorphism. T is a covariant additive functor of the category of Z-modules in the category of R-modules. With T is associated the covariant functor T' of the category of Z-chain complexes into the category of R-chain complexes. The space pair (X, A) then has an associated R-chain complex consisting of the R-modules

$$S_q(X, A) = T(S_q(X, A)) = S_q(X, A) \otimes_Z M,$$

and the boundary operators $\partial_q \otimes_Z 1$.

Clearly, $S_q(X, A) = \sum{}^{\oplus} Z_\sigma$, and therefore, by Theorem 9.4, $\bar{S}_q(X, A) = \sum{}^{\oplus}(Z_\sigma) \otimes_Z M$, where the sums are taken over all q-simplices σ with $|\sigma| \subset X$, $|\sigma| \not\subset A$. Here, it follows from 8.7 that $Z_\sigma \cong Z$ and $(Z_\sigma) \otimes_Z M \cong M$. The elements of $\bar{S}_q(X, A)$ can therefore be written as finite sums of the form $\sum m_\sigma \sigma$, where $m_\sigma \sigma$ is defined by $m_\sigma \sigma = \sigma \otimes_Z m_\sigma$ for $m_\sigma \in M$. It is clear that computations are to be made with these sums according to the usual rules for sums and products. The boundaries of such elements are found from

$$(\partial \otimes_Z 1)(\sum \sigma \otimes_Z m_\sigma) = \sum ((\partial\sigma) \otimes_Z m_\sigma) = \sum m_\sigma \partial\sigma$$

and are in $\bar{S}_{q-1}(X, A)$. Corresponding to a continuous $f:(X, A) \longrightarrow (Y, B)$ there is an $f^\#$ and a $T(f^\#) = f \otimes_Z 1$. For these, the definition of $f^\#$ yields

$$(f^\# \otimes_Z 1)(\sum m_\sigma \sigma) = \sum (f^\# \sigma \otimes m_\sigma) = \sum m_\sigma f(\sigma).$$

This proves that *the old singular theory of* Chapters 4–7 *still holds if R is replaced by an R-module M*. The choice $M = R$ shows that it would have been sufficient to develop the old theory for Z. However, no simplifications would have occurred thereby.

Great care must be exercised in the transition from the results of the old theory. It is worth heeding the **warning:** *In general, $H_q^Z(X, A) \otimes_Z M$ is not isomorphic to $H_q^M(X, A)$.* The details can be seen in the book, *Homology Theory* by S. McLane, under the heading of "Universal Coefficient Theorems."

A review of the proofs in Chapter 7 will yield the following:

11.1. Direct Decomposition Theorem: *Let $\{X_k\}$ be the set of path-components of X; let $A \subset X$ and $A_k = A \cap X_k$.* Then, for non-augmented homology,

$$\sum^{\oplus} H_q(X_k, A_k) = H_q(X, A).$$

11.2. Computation of $H_0(X, A)$: *Let r $(0 \leq r \leq \infty)$ be the cardinality of the path-components X_k of X for which $X_k \cap A = \varnothing$. Then $H_0(X, A) = \sum^{\oplus} M$, where, in the augmented homology, the number of summands is r when $A \neq \varnothing$ and $r - 1$ when $A = \varnothing$. In the non-augmented homology, the number is always r.*

11.3. Homology Groups of Points: *In the non-augmented case, $H_0(P) = M$ and $H_q(P) = 0$ when $q \neq 0$.*

11.4. Homology Groups of S^n: *In the augmented theory, $H_q(S^n) = 0$ for $q \neq n$ and $H_n(S^n) \cong M$.*

Cohomology

For the functor T choose $T = \text{Hom}(\ \ , M)$; that is, if C is a Z-module, then $T(C) = \text{Hom}_Z(C, M)$. To a Z-homomorphism $f: C \longrightarrow D$ there corresponds $T(f) = \text{Hom}_Z(f, 1)$: $\text{Hom}_Z(D, M) \longrightarrow \text{Hom}_Z(C, M)$ where $\text{Hom}_Z(f, 1)(\phi) = \phi f$. Since M is an R, Z-module, $T(C)$ is an R-module and $T(f)$ is an R-homomorphism. By Chapter 9, T is a contravariant additive functor of the Z-modules into the R-modules. With T is associated the contravariant functor T' of the Z-chain complexes in the R-cochain complexes. The space pair (X, A) then has an associated cochain complex

$$\cdots \longleftarrow S^{q+1}(X, A) \xleftarrow{\delta^{q+1}} S^q(X, A) \xleftarrow{\delta^q} S^{q-1}(X, A) \longleftarrow \cdots,$$

consisting of the R-modules

$$S^q(X, A) = T(S_q(X, A)) = \text{Hom}_Z(S_q(X, A), M)$$

and the coboundary operators $\delta^q = \text{Hom}_Z(\partial_q, 1)$. The elements of $S^q(X, A)$ are called q-cochains, those of Ker δ^{q+1} are called q-cocycles, and those of Im δ^q are called q-coboundaries. The factor group of the q-cocycles modulo the q-coboundaries is called the qth cohomology group $H^q(X, A)$.

A q-cochain is a Z-homomorphism $\phi: S_q(X, A) \longrightarrow M$. A q-cochain is determined by its values on a generating system. Since $S_q(X, A)$ is free, a q-cochain is obtained when arbitrary values of M are assigned to the elements of a basis. For $q \geq 0$, the q-cochains; i.e., the elements $\phi: S_q(X, A) \longrightarrow M$, can be interpreted as arbitrary functions of the generating q-simplices σ with values in M. This function value will be written $\phi(\sigma)$. For $\psi \in S^{q-1}(X, A)$, it is easy to see that $\delta^q\psi = \text{Hom}_Z(\partial, 1)(\psi) = \psi\partial_q$ where $\psi\partial_q \in S^q(X, A)$. It has then been proved that

11.5. $$(\delta^q\psi)(\sigma) = \psi(\partial_q\sigma),$$

where $\partial_q\sigma$ is the familiar boundary, a linear combination of $(q-1)$-simplices.

Direct Decomposition: Let (X, A) be a space pair. Let $\{X_k\}$ be the set of path-components of X and let $A_k = A \cap X_k$. Just preceding Theorem 7.2, a chain isomorphism $\sum i_k^\#: \sum^\oplus S_q(X_k, A_k) \longrightarrow S_q(X, A)$ was associated with the injections $i_k: (X_k, A_k) \longrightarrow (X, A)$. Theorem 9.3 shows that there is a chain isomorphism $S^q(X, A) \longrightarrow \prod S^q(X_k, A_k)$. From this follows

11.6. THEOREM: *The induced homomorphisms $i_k^*: H^q(X, A) \longrightarrow H^q(X_k, A_k)$ form a projective representation of $H^q(X, A)$ as a direct product (in the non-augmented case).*

Computation of $H^0(X, A)$: Let X be pathwise-connected (for example, let X be a region in the complex plane and let M be the complex numbers). A 0-cochain ψ is an arbitrary function on the 0-simplices with values in M. Its coboundary $\delta\psi$ is a function on the 1-simplices; that is, the paths σ in X. Let P be the initial point and Q be the end point of σ. Then

$$(\delta^1\psi)(\sigma) = \psi(\partial_1\sigma) = \psi(Q) - \psi(P).$$

Since X is pathwise-connected, it follows that $\psi \in \text{Ker}\,\delta^1$ is equivalent to $\psi(x) = \text{constant}$.

In the non-augmented theory, $S_{-1}(X) = 0$ and therefore, $S^{-1}(X) = 0$. For $A \neq 0$, it follows that $\delta^0(S^{-1}(X)) = 0$, and then that

$$H^0(X) = \text{Ker } \delta^1/\text{Im } \delta^0 \cong M.$$

Now let $A \subset X$, where $A \neq 0$. Let P be a point of A. To each point $Q \in X$ there is a path from P to Q, say σ. If ψ is a 0-cochain, then $(\delta^1\psi)(\sigma) = \psi(Q)$ since $P \in A$, and therefore, $(P) \in S_q(A)$. It follows from $\psi \in \text{Ker } \delta^1$ that $\psi Q = 0$ for each $Q \in X$. Therefore, $\psi = 0$ and $H^0(X, A) = 0$.

11.7. Cohomology Groups of a Point P: It was just proved that $H^0(P) \cong M$ in the non-augmented case. In the augmented homology, $\partial_0\sigma_0 = (.)$ and then $(\delta^0\phi)(\sigma_0) = \phi(.)$. Therefore, δ^0 is an isomorphism. This means that all 0-cochains are coboundaries, and hence, $H^0(P) = 0$. If $q < -1$, the cohomology groups are zero by definition. When $q = -1$, the fact that $\delta^0 : S^{-1}(P) \longrightarrow S^0(P)$ is an isomorphism means that Ker $\delta^0 = 0$, and hence, that $H^{-1}(P) = 0$. Now let $q > 0$. By the proof of 7.5, $\partial_q\sigma_q = 0$ when q is an odd integer, and $\partial_q\sigma_q = \sigma_{q-1}$ when q is even where $\sigma_q : \Delta_q \longrightarrow P$. When $q > 0$ is odd, it follows that

$$(\delta^q\phi)\sigma_q = \phi(\partial_q\sigma_q) = 0,$$

and hence, that δ^q is the 0-homomorphism. When $q > 0$ is even, it follows that $(\delta^q\phi)(\sigma_q) = \phi(\sigma_{q-1})$; i.e., that δ^q is an isomorphism. When $q > 0$ is even, it then follows that

$$H^q(P) = (\text{Ker } \delta^{q+1})/(\text{Im } \delta)^q \cong M/M = 0.$$

When $q > 0$ is odd, it follows that $H^q(P) = 0/0 = 0$. Thus, $H^q(P) = 0$ when $q \neq 0$.

11.8. The Cohomology Groups of S^n: The proof of 7.12 remains valid except for the direction of the arrows. This means, in the augmented case, that $H^n(S^n) \cong M$ and $H^q(S^n) = 0$ when $q \neq n$. In the non-augmented case, $H^0(S^0) \cong M + M$; $H^0(S^n) \cong M$ and $H^n(S^n) \cong M$ when $n > 0$; and $H^q(S^n) = 0$ when $q \neq 0, q \neq n$.

Application of Homology to Function Theory

The *argument* of a complex number $z = r(\cos \alpha + i \sin \alpha) \neq 0$ is a name of the angle $\alpha = \arg z$. It is only determined up to a multiple of 2π. A line g in the complex plane C defines two half-planes H and H_1. Choose $z' \in H$ and $z_1' \in H_1$ so that the vector $z_1' - z'$ is perpendicular to g. The choice of a fixed value β for $\arg(z_1' - z')$ permits the determination of the value of $\arg(z_1 - z)$, $z_1 \in H_1$, $z \in H$ by means of the statement

$$\beta - \pi/2 < \arg(z_1 - z) < \beta + \pi/2.$$

Then $\arg(z_1 - z)$ is a continuous function of $z \in H$ and $z_1 \in H_1$.

The Winding Number: Let $\sigma: I \longrightarrow X$ be a (continuous) 1-simplex. Let z_0 be a point not in $|\sigma|$ whose distance from $|\sigma|$ is ρ. Since $|\sigma|$ is compact, there are numbers $\{t_i\}$ such that $0 = t_0 \leq t_1 \leq \cdots \leq t_n = 1$ for which the variation of $\sigma(t)$ on each subinterval $I_j = \{t \mid t \in I, t_{j-1} \leq t \leq t_j\}$ is less than $\rho/2$. Choose a point $\sigma(t_j) \in I_j$. Erect on the line segment from z_0 to $\sigma(t_j)$ a perpendicular line g_j whose distance from z_0 is $\rho/2$. Then the points $\sigma(t)$, where $t \in I$, and the points z, for which $|z - z_0| < \rho/2$, lie in distinct half-planes with respect to g_j. Therefore, when $t \in I_j$ and z satisfies $|z - z_0| < \rho/2$, there is a continuous function of t and z, say $a_j(t, z)$, whose value at (t, z) is an argument of $\sigma(t) - z$. By addition of a suitable multiple of 2π, it is possible to arrange that $\alpha_{j+1}(t_j, z) = \alpha_j(t_j, z)$, so that there exists a composite function $\alpha(\sigma, t, z)$ which is continuous in (t, z) when $t \in I$ and $|z - z_0| < \rho/2$, and whose value is an argument of $\sigma(t) - z$.

For a second function $\beta(\sigma, t, z)$ with the same properties, $\alpha(\sigma,t,z) - \beta(\sigma,t,z)$ is an integer multiple of 2π. Since the difference is continuous in t, it must be constant. Consequently, $\alpha(\sigma,t,z)$ is uniquely determined up to a constant $2\pi n$. Denote the uniquely determined value of $\alpha(-, 1, z) - \alpha(-, 0, z)$ by $V(\sigma, z)$. This is called the *variation in the argument of σ with respect to z*, and is a continuous function of z. $V(\sigma, z)$ is only weakly dependent upon the parametrization of the set $|\sigma|$. To see this, let $h: I \longrightarrow I$ be a continuous, strictly monotone, increasing function. Then $V(\sigma h, z) = V(\sigma, z)$, because $\alpha(\sigma, h(t), z)$ can be taken to be $\alpha(\sigma h, t, z)$.

Let $0 = t_0 < t_1 < \cdots < t_n = 1$ be a subdivision of I into intervals $I_j = \{t \mid t \in I, t_{j-1} \leq t \leq t_j\}$. Choose a continuous, strictly monotone, increasing function h_j of I on I_j and let $\sigma_j = \sigma h_j$. The simplices σ_j will be called a *subdivision* of σ. For the function $\alpha(\sigma_j, t, z)$ it is possible to use $\alpha(\sigma, h_j(t), z)$. Therefore,

$$\begin{aligned} V(\sigma, z) = \alpha(\sigma, 1, z) - \alpha(\sigma, 0, z) &= \sum_{j=1}^{n} (\alpha(\sigma, t_j, z) - \alpha(\sigma, t_{j-1}, z)) \\ &= \sum_{j=1}^{n} V(\sigma_j, z). \end{aligned}$$

The $V(\sigma_j, z)$ are not dependent upon the particular choice of h_j since for another h_j, say h'_j, $h_j^{-1}h'_j: I \longrightarrow I$ is a strictly monotone, increasing function.

Let A be a closed set disjoint from $|\sigma|$, and let $\rho > 0$ be its distance from $|\sigma|$. Let $\{\sigma_j\}$ be a subdivision of σ such that each $|\sigma_j|$ has diameter $< \rho/2$. If $z \in A$, $V(\sigma_j, z)$ depends only upon the points $\sigma(t_j)$ and $\sigma(t_{j-1})$ and is less than π. The line segment connecting $\sigma(t_j)$ and $\sigma(t_{j-1})$ is the carrier of a 1-simplex γ_j. Clearly, $V(\gamma_j, z) = V(\sigma_j, z)$. Hence, if γ is the polygonal line formed by assembling the γ_j,

$$V(\gamma, z) = \sum V(\gamma_j, z) = \sum V(\sigma_j, z) = V(\sigma, z).$$

Thus, the variation in the argument of σ is the same as that of a sufficiently fine inscribed polygon.

If c is a 1-chain, $c = m_1\sigma_1 + \cdots + m_r\sigma_r$, $m_i \in Z$, then $V(c, z)$ is defined by

$$V(c, z) = m_1V(\sigma_1, z) + \cdots + m_rV(\sigma_r, z)$$

when $z \notin |c|$. If P_i is the initial point, and Q_i is the terminal point of σ_i, then

$$V(c, z) = \sum m_i(\arg(Q_i - z) - \arg(P_i - z))$$

up to a multiple of 2π. Now let c be a cycle. Then

$$\partial c = \sum m_i((Q_i) - (P_i)) = 0.$$

Thus, if the weights are counted, each end point is as often a Q_i as it is a P_i. Hence, $V(c, z) = 0$ up to a multiple of 2π. If c is a cycle, the integer

$$W(c, z) = \frac{1}{2\pi} V(c, z)$$

is called the *winding number* of c with respect to z. If a cycle is replaced by the cycle formed by a sufficiently fine inscribed polygon, then the winding number is not changed.

Now let c be a cycle whose carrier consists of s line segments. Each segment defines a line, and, thereby, two open half-planes. A point which does not lie on any of the lines is in s of the half-planes. Their intersection, being an intersection of convex sets, is convex, and is a polygon in case it is bounded. Cut this intersection into triangles and a possible unbounded region by the diagonals from one of the vertices. The plane is then divided into a definite set of triangles and unbounded regions by means of the lines and the chosen diagonals. Let $\mathscr{M}$ be one of the triangles or unbounded regions. Since $\mathscr{M}$ is connected, and the integer $W(c, z)$ is a continuous function on $\mathscr{M}$, it is constant there. Let $\mathscr{M}$ be unbounded. When $z \in \mathscr{M}$ and $|z|$ is large, then $V(\sigma, z)$ is very small. Therefore, $W(c, z)$ is very small, and is consequently equal to 0. Let $\mathscr{M}_i$, $i = 1, 2, \cdots, r$, be the triangular regions for which $W(c, z) = w_i \neq 0$ for $z \in \mathscr{M}_i$. Choose a 2-simplex Δ_i for which $\mathscr{M}_i$ is the interior of $|\Delta_i|$. The orientation of Δ_i is to be chosen in such a manner that $W(\partial\Delta_i, z) = 1$ for $z \in \mathscr{M}_i$.

Now form the new chain $c' = c - w_1\partial\Delta_1 - \cdots - w_r\partial\Delta_r$. That $W(c', z) = 0$ for all z which are not on one of the boundaries of the decomposition of the plane can be seen as follows: If z lies in none of the $|\Delta_i|$, then $W(c, z) = 0$ and $W(\partial\Delta_i, z) = 0$, and therefore $W(c', z) = 0$. If z lies in the interior of $|\Delta_i|$, then $W(c, z) = w_i$, $W(\partial\Delta_i, z) = 1$, and $W(\partial\Delta_j, z) = 0$ for $j \neq i$. Hence, $W(c', z) = 0$.

The carrier of c' consists of line segments. Subdivision of c' yields a chain c'' such that the interior of each line segment is free of end points of other segments. Let s_i, $i = 1, 2, \cdots, m$ be the distinct segments of c''. Let n_i be the sum of the weights with which the simplices having s_i as carrier occur. Choose points z_1 and z_2 very near an inner point of s_1 but on opposite sides of s_1 in such a manner that the segment connecting them meets only s_1 and none of the other s_i. Then $V(s_i, z_1) - V(s_i, z_2)$ is very small for $i \neq 1$ and near $\pm 2\pi$ for $i = 1$. Therefore, $W(c_1'' z_1) - W(c_1'' z_2)$ is approximately $\pm n_1$. It was previously shown that $W(c_1'' z_1) = 0 = W(c_1'' z_2)$. Therefore, $n_1 = 0$, and similarly, $n_2 = \cdots = n_m = 0$. It has therefore been proved that each segment of c'' is traversed just as often in the one direction as the other.

Now let X be a preassigned open subset of the plane. A *chain of* X shall mean a chain c for which $|c| \subset X$. A cycle c of X is called *homologous to* 0 (in symbols, $c \sim 0$) if $W(c, z) = 0$ for every point z of the complement of X. Suppose that $c \sim 0$. Then each sufficiently fine polygonal cycle inscribed in c is homologous to zero. If c' arises from c by subdivision or by a reparametrization, described before, then $c' \sim 0$ also. Now let c be a cycle whose carrier consists of line segments and let $c \sim 0$. Consider the corresponding triangles Δ_i. Let $z \in |\Delta_i|$. If $z \in |c|$, then $z \in X$; otherwise, $W(c, z)$ is defined and constant in a neighborhood of z. This neighborhood also contains inner points of $|\Delta_i|$. Hence, $W(c, z) = w_i \neq 0$. From $c \sim 0$, it follows that $z \in X$. Therefore, $|\Delta_i| \subset X$.

Let c be a chain of X with a polygonal carrier and let $c \sim 0$. Then there is a representation $c = c' + w_1\partial\Delta_1 + \cdots + w_r\partial\Delta_r$, where $\Delta_1, \cdots, \Delta_r$, and c' are chains of X. After a subdivision, each segment of c' is traversed as often in one direction as the other. Hence, c' is a boundary, and c is also a boundary. Conversely, each boundary is homologous to zero, as is easy to see. This justifies the definition of the word homologous.

Integration: Let $f(z)$ be a continuous, complex-valued function defined on X. A 1-chain is called *rectifiable* or *a path of integration* if and only if the carrier of its simplices is rectifiable. A line integral $\int_\sigma f(z)dz$ is associated with each 1-simplex σ of X. If the path of integration is $c = \sum m_i\sigma_i$, the integral $\int_c f(z)dz$ is defined by $\int_c f(z)dz = \sum m_i \int_{\sigma_i} f(z)dz$. This integral is not changed by subdivision of c or by monotone changes of the parametrization. A path of integration is called *closed* if c is a cycle.

11.9. CAUCHY INTEGRAL THEOREM: *Let $f(z)$ be analytic in X and let c be a closed path of integration in X that is homologous to zero. Then*

$$\int_c f(z)\,dz = 0.$$

Proof: Let Δ be a 2-simplex whose carrier is a triangle contained in X. Many books on function theory show that $\int_{\partial\Delta} f(z)dz = 0$. Now suppose that c is a cycle of X, and that $c \sim 0$. The value of the integral is the limit of the integrals over sufficiently fine polygonal cycles that are inscribed in c. Therefore it is sufficient to assume c to be polygonal and homologous to zero. From the representation $c = c' + \sum w_i\partial\Delta_i$ and $\int_{\partial\Delta_i} f(z)dz = 0$, it follows that $\int_c f(z)dz = \int_{c'} f(z)dz$. In a suitable subdivision of c', all the segments are traversed equally often in both directions. Hence, $\int_{c'} f(z)dz = 0$.

Consider as an example the case $f(z) = 1/z$. This function is analytic for $z \neq 0$. Let X consist of the plane C from which the origin has been deleted. Let K be the unit circle parametrized by

$\cos(2\pi t) + i \sin(2\pi t)$, $t \in I$. The value of $\alpha(k, t, 0)$ can be taken to be $2\pi t$, from which it follows that $W(K, 0) = 1$. Let c be an arbitrary cycle of X and $W(c, 0) = w$. Then $c - wK = c'$ is a cycle for which $W(c', 0) = 0$ and $c' \sim 0$. Therefore, $\int_{c'} (1/z)\,dz = 0$. Hence,

$$\int_c \frac{1}{z}\,dz = w \int_K \frac{1}{z}\,dz = 2\pi i W(c, 0).$$

Cohomology: From the de Rham Theorem (not to be proved here) it follows that the cohomology theory is not affected if considerations are restricted to rectifiable 1-simplices. Instead of the set of all complex-valued functions defined on the chains, it is sufficient to deal only with those functions $\phi(\sigma)$ that can be written $\phi(\sigma) = \int_\sigma f(z)\,dz$ where f is continuous on X. The function f is uniquely determined by ϕ. A computation of $H^1(X) = \text{Ker } \delta^2/\text{Im } \delta^1$ will now be carried out: The statement, $\phi \in \text{Ker } \delta^2$ is equivalent to $\delta^2\phi(\tau) = \phi(\partial\tau) = 0$ for all triangles τ with rectifiable sides. This is equivalent to $\int_{\partial\tau} f(z)\,dz = 0$ and thus to the analyticity of f on X by the Theorem of Morera. Hence, $\phi \in \text{Ker } \delta^2$ if and only if $f(z)$ is analytic in X. A $\psi \in S^0(X)$ assigns to each point P of X [actually to each 0-simplex (P)] the number $\psi(P)$. Let σ be a rectifiable 1-simplex with initial point P and end point Q. Then $(\delta^1\psi)(\sigma) = \psi(\partial_1\sigma) = \psi(Q) - \psi(P)$. Hence, $\phi \in \delta^1(S^0(X))$ if and only if there is a $\psi \in S^0(X)$ such that $\int_\sigma f(z)\,dz = \psi(Q) - \psi(P)$; that is, if the integral is independent of the path connecting the points P and Q in X. This in turn means that $\int_c f(z)\,dz = 0$ for all closed paths of integration.

Let $\phi \in \text{Ker } \delta^2$; i.e., let $f(z)$ be an analytic function on X. To each closed path of integration c associate a complex number $\text{Res}_f(c) = (1/2\pi i)\int_c f(z)\,dz$. The function Res_f is called the *residue function* of f. If g is also analytic and $\text{Res}_g = \text{Res}_f$, then $\text{Res}_{f-g}(c) = 0$ for all closed paths c in X, and then $\phi = \int f(z)\,dz$ and $\chi = \int g(z)\,dz$ differ only by a coboundary $\delta^1\psi$. Conversely, if ϕ and χ differ only by a coboundary, then $\text{Res}_f(c) = \text{Res}_g(c)$ for all c. Therefore, *the group $H^1(X)$ is isomorphic to the set of residue functions.*

Remove a collection of r compact connected subsets from a simply connected open subset of the complex plane. The resulting region X is said to have r holes. In each hole choose a point, and let x_k be the chosen point in the kth hole. Let c_k be a closed path of integration in X which has the winding number 1 with respect to x_k and 0 with respect to x_j for $j \neq k$. Let c be an arbitrary closed path of integration in X with the winding number w_i about x_i. For $c' = c - w_1 c_1 - \cdots - w_r c_r$ it follows that $W(c', z_0) = 0$ for each $z_0 \notin X$. From the Cauchy Integral Theorem it follows that

$$\int_c f(z)dz = w_1 \int_{c_1} f(z)dz + \cdots + w_r \int_{c_r} f(z)dz.$$

Consequently, $\text{Res}_f(c)$ is completely described by the r numbers $\alpha_k = \int_{c_k} f(z)dz$. Conversely, to each set of r arbitrary complex numbers α_k there corresponds the function $g(z) = (1/2\pi i)\sum(\alpha_k)/(z - x_k)$ which is analytic in X and for which $\int_{c_k} g(z)dz = \alpha_k$. Thus, $H^1(X)$ is isomorphic to the direct sum of r summands C.

The relations 7.18 say the same about $H_1(X)$. The proof at that point can be carried over here, for X is homotopic to a segment with r attached circles, and the cohomology groups of a circle are known by 11.8.

12

Axioms for Homology and Cohomology

In addition to the singular theory, there are other theories for homology and cohomology. The following axioms (of Eilenberg and Steenrod) are weakened theorems of the singular theory which also hold in other theories.

Let $\mathscr{K}$ be a category of certain space pairs and continuous functions satisfying the following conditions:

(1) If $(X, A) \in \mathscr{K}$, then the pairs $(X, \varnothing) = X, (A, \varnothing) = A, (X, X), (A, A)$, and $(\varnothing, \varnothing)$, together with all injections between them, lie in $\mathscr{K}$.

(2) If $(f:(X, A) \longrightarrow (Y, B)) \in \mathscr{K}$, then $\mathscr{K}$ contains all the functions induced by f on the pairs mentioned in (1).

(3) If $(X, A) \in \mathscr{K}$, then $\mathscr{K}$ contains $(X \times I, A \times I)$ and the two functions defined by $\lambda(x) = (x, 0)$ and $\mu(x) = (x, 1)$.

(4) $\mathscr{K}$ contains a space P_0 consisting of a single point. If $X \in \mathscr{K}$ and P is a singleton space in $\mathscr{K}$, then each function $P \longrightarrow X$ is in $\mathscr{K}$.

The space pairs and functions in $\mathscr{K}$ are called *admissible*. A space triple (X, A, B) is called admissible if and only if (X, A), (Y, B), and (A, B) are admissible space pairs.

Examples: Until now, consideration has been given to the set of all space pairs and continuous functions. They satisfy the conditions. Another example is the category of all pairs (X, A), in which X and A are compact hausdorff spaces, together with the continuous functions between these pairs.

A homology (cohomology) theory on $\mathscr{K}$ associates with each ring R, each space pair $(X, A) \in \mathscr{K}$, and each integer q an R-module $H_q(X, A)$ $(H^q(X, A))$. Each continuous function $f \in \mathscr{K}$, $f:(X, A) \longrightarrow (Y, B)$ has an associated R-homomorphism $f_*:H_q(X, A) \longrightarrow H_q(Y, B)$ $(f^*:H^q(Y, B) \longrightarrow H^q(X, A))$. Furthermore, there is to each admissible space triple (X, A, B) on R-homomorphism $\partial_*:H_q(X, A) \longrightarrow H_{q-1}(A, B)$ $(\delta^*:H^{q-1}(A, B) \longrightarrow H^q(X, A))$. For these homomorphisms, the following seven axioms must be satisfied:

I. $\mathrm{Id}_* = \mathrm{Id}$ $(\mathrm{Id}^* = \mathrm{Id})$.

II. $(gf)_* = g_*f_*$ $((gf)^* = f^*g^*)$.

The first two axioms state that homology theories are covariant functors and cohomology theories are contravariant functors.

III. *If* $f:(X, A, B) \longrightarrow (X', A', B')$, *then* ∂_* (δ^*) *commutes with the induced homomorphisms of the homology (cohomology) groups of the space pairs.*

IV. *The sequence*

$$\cdots \longrightarrow H_q(A, B) \xrightarrow{i_*} H_q(X, B) \xrightarrow{j_*} H_q(X, A) \xrightarrow{\partial_*} H_q(A, B) \cdots$$

$$(\cdots \longleftarrow H^q(A, B) \xleftarrow{i^*} H^q(X, B) \xleftarrow{j^*} H^q(X, A) \xleftarrow{\delta^*} H^q(A, B) \longleftarrow \cdots)$$

of homology (cohomology) groups is exact. These are the familiar sequences.

For an admissible $f:(X, A, B) \longrightarrow (X', A', B')$, it follows from the first three axioms that all rectangles of the diagram

$$\begin{array}{ccccccccc}
\cdots \longrightarrow H_q(A,B) & \xrightarrow{i_*} & H_q(X,B) & \xrightarrow{j_*} & H_q(X,A) & \xrightarrow{\partial_*} & H_q(A,B) & \longrightarrow \cdots \\
\downarrow & & \downarrow & & \downarrow & & \downarrow & \\
\cdots \longrightarrow H_q(A',B') & \xrightarrow{i'_*} & H_q(X',B') & \xrightarrow{j'_*} & H_q(X',A') & \xrightarrow{\partial_*} & H_q(A',B') & \longrightarrow \cdots
\end{array}$$

are commutative in a homology theory (the analogous diagram for cohomology theory is also commutative).

V. *Homotopy Axiom. For the functions* $\lambda,\mu\colon (X, A) \longrightarrow (X \times I, A \times I)$ *given by* $\lambda(x) = (x, 0)$ *and* $\mu(x) = (x, 1)$, *the equality* $\lambda_* = \mu_*$ $(\lambda^* = \mu^*)$ *holds.*

VI. *Excision Axiom. Let U be an open subset of X for which* $\overline{U} \subset \mathring{A}$. *If the injection* $i\colon(X - U, A - U) \longrightarrow (X, A)$ *is admissible, then* $i_*(i^*)$ *is an isomorphism.*

VII. *Dimension Axiom. If P consists of a single point and* $P \in \mathscr{K}$, *then* $H_q(P) = 0$ $(H^q(P) = 0)$ *for* $q \neq 0$.

Obviously, the non-augmented singular theory with coefficients in M satisfies all the axioms. The excision axiom was even proved without the hypothesis that U is open.

Let $P \in \mathscr{K}$ be a singleton space. The module $H_0(P)$ $(H^0(P))$ is determined up to an isomorphism and is called the *coefficient module* of the theory.

Let (X, X) be admissible. By Axiom IV, the sequence

$$\cdots \longrightarrow H_q(X) \xrightarrow{i_*} H_q(X) \xrightarrow{j_*} H_q(X, X) \xrightarrow{\partial_*} H_{q-1}(X) \longrightarrow \cdots$$

is exact. Since $i = \text{Id}$, $i_* = \text{Id}$. Exactness then yields $j_* = 0$ and $\partial_* = 0$. Therefore, $H_q(X, X) = 0$. Analogously, $H^q(X, X) = 0$ in a cohomology theory. In particular, $H_q(P_0, P_0) = 0$ or $H^q(P_0, P_0) = 0$, respectively. The axioms do *not* yield $H_q(X) = 0$ for $q < 0$. However, this can be proved for sufficiently simple spaces. (Examples will follow.)

The stated axioms are those of a non-augmented theory. Corresponding to this, there is an augmented theory which will be constructed now so that both can be used in examples.

Definition: Let M_1 and M_2 be modules, and let $f\colon M_1 \longrightarrow M_2$ be a homomorphism. Then $\operatorname{Coker} f = M_2/\operatorname{Im} f$.

12.1. Lemma: *Let C and C′ be modules and* $p\colon C \longrightarrow C'$ *and* $\psi\colon C' \longrightarrow C$ *be homomorphisms with* $p\psi = \text{Id}$. *Then* ψ *is a monomorphism, p is an epimorphism, and*

12.2. $\qquad C = \operatorname{Ker} p \oplus \operatorname{Im} \psi = \operatorname{Coker} \psi \oplus \operatorname{Im} \psi.$

Proof: Let $i{:}\operatorname{Ker} p \longrightarrow C$ be the injection. If λ' is the function given by $\lambda' = \mathrm{Id} - \psi p{:}C \longrightarrow C$, then $p\lambda' = p - p\psi p = 0$. Therefore, it is possible to define a homomorphism $\lambda{:}C \longrightarrow \operatorname{Ker} p$ by $\lambda(x) = \lambda'(x)$. Since $\lambda'\psi = \psi - \psi p \psi = 0$, it follows that $\lambda\psi = 0$. Hence, the assumptions of Lemmas 10.4 and 10.5 are satisfied for the diagram

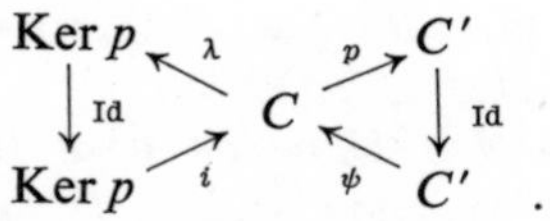

This yields an injective and a projective representation of C as a direct sum of C and $\operatorname{Ker} p$. By Lemma 10.5, the diagonal from C' to $\operatorname{Ker} p$ is exact. Since ψ is a monomorphism and λ is an epimorphism, it follows that $\operatorname{Ker} p = C/\psi(C') = \operatorname{Coker} \psi$.

12.3. THEOREM: *Let* $p{:}C \longrightarrow C'$ *and* $k{:}C' \longrightarrow C$ *be chain homomorphisms with* $pk = \mathrm{Id}$. *Let* $A_q = \operatorname{Ker} p_q$ *and* $B_q = k_q(C_q')$ *be the direct summands of* C_q *found in Lemma* 12.1. *Let* C *be exact with boundary operator* α. *Then* A *and* B *are exact chain complexes under the restrictions of* α.

Proof: Designate the boundary operator of C' by β. Since

$$p(\alpha(A_q)) = \beta(p(A_q)) = 0,$$

it follows that

$$\alpha(A_q) \subset \operatorname{Ker} p_{q-1} = A_{q-1}.$$

Hence, α induces a homomorphism $\bar{\alpha}{:}A_q \longrightarrow A_{q-1}$. Similarly, it can be proved that B is a chain complex. Now let C be exact. Then to each $a \in A_q$ for which $\bar{\alpha}(a) = 0$, and hence, $\alpha(a) = 0$, there is a $c \in C_{q+1}$ such that $\alpha(c) = a$. However, $c = a' + b$ where $a' \in A_{q+1}$ and $b \in B_{q+1}$. Since $\alpha(c) = a$ and $\alpha(a') \in A_q$, it follows that $\alpha(b) \in A_q \cap B_q$, that $\alpha(b) = 0$, and then that $\alpha(a') = \alpha(c) = a$. Therefore, A is exact. The exactness of B is proved analogously.

12.4. Corollary: *If* $p{:}(X, A) \longrightarrow (Y, B)$ *and* $k{:}(Y, B) \longrightarrow (X, A)$ *are admissible and* $pk = \mathrm{Id}$, *then the exact homology sequence*

$$\cdots \longrightarrow H_q(X, A) \xrightarrow{\partial_*} H_q(A) \xrightarrow{i_*} H_q(X) \xrightarrow{j_*} H_q(X, A) \longrightarrow \cdots$$

splits into two exact sequences: The one consisting of the kernels of p_*, and the other consisting of the images of k_*. The exact cohomology sequence splits into two sequences: One consisting of the cokernels of k^* and the other of the images of p^*.

Proof: From $pk = \text{Id}$, it follows that $p_*k_* = \text{Id}$ or $k^*p^* = \text{Id}$, as the case may be.

Application: Let $P \in \mathscr{K}$ be a singleton space. For each space pair (X, A), the function $k:(P, P) \longrightarrow (X, A)$ is admissible by (1) and (2). Now let the function $p:(X, A) \longrightarrow (P, P)$ be admissible. Since $pk = \text{Id}$, the corollary can be applied. Since $H_q(P, P) = 0$ for all q, since $H_0(P) = M$ and since $H_q(P) = 0$ for $q \neq 0$, the exact sequence of Ker p_* is the homology sequence of (X, A) except for two locations. For $p:X \longrightarrow P$, let Ker $p_* = \tilde{H}_0(x)$. This yields the so-called *reduced homology sequence*

$$\cdots \longrightarrow H_1(X,A) \xrightarrow{\bar{\partial}_*} \tilde{H}_0(A) \xrightarrow{\bar{i}_*} \tilde{H}_0(X) \xrightarrow{\bar{j}_*} H_0(X,A) \xrightarrow{\partial_*} H_{-1}(A) \longrightarrow \cdots.$$

In cohomology, $\tilde{H}^0(X)$ is defined to be Coker k^* where $k:P \longrightarrow X$; and the corollary yields the *reduced cohomology sequence:*

$$\cdots \longleftarrow H'(X,A) \xleftarrow{\delta^*} \tilde{H}^0(A) \xleftarrow{i^*} \tilde{H}^0(X) \xleftarrow{j^*} H^0(X,A) \xleftarrow{\delta^*} H^{-1}(A) \longleftarrow \cdots.$$

The groups $\tilde{H}_0(X)$ and $\tilde{H}^0(X)$ are called the *reduced* 0- *dimensional homology* and *cohomology groups*, respectively. Clearly

$$H_0(X) \cong \tilde{H}_0(X) + k_*(M)$$

and

$$H^0(X) \cong \tilde{H}^0(X) + p^*(M).$$

12.5. THEOREM: *Let $f:X \longrightarrow Y$ and $p':Y \longrightarrow P$ be admissible. Then $p = p'f$ is admissible and the homomorphisms $f_*:H_0(X) \longrightarrow H_0(Y)$ and $f^*:H^0(Y) \longrightarrow H^0(X)$ induce homomorphisms $\tilde{f}_*:\tilde{H}^0(X) \longrightarrow \tilde{H}_0(Y)$ and $\tilde{f}^*:\tilde{H}^0(Y) \longrightarrow \tilde{H}^0(X)$, respectively.*

Proof: Let $c \in \tilde{H}_0(X) = \text{Ker } p_*$. Then $0 = p_*(c) = p'_*f_*(c)$ and $f_*(c) \in \tilde{H}_0(Y)$. Since $f^*p'^*(M) \subset p'(M)$, Lemma 1.1 shows that f^*

induces a homomorphism of $\tilde{H}^0(Y) = H^0(Y)/p'^*(M)$ into $\tilde{H}^0(X) = H^0(X)/p^*(M)$.

Thus, if the category $\mathscr{K}$ is restricted to those space pairs for which the function $(X, A) \longrightarrow (P, P)$ is admissible, an augmented theory has been obtained for the restricted category. In this theory, $\tilde{H}_0(P) = 0$ and $\tilde{H}^0(P) = 0$ since, for $p = k = \text{Id}\colon P \longrightarrow P$, both Ker $p_* = 0$ and Coker $p^* = 0$.

In the non-augmented singular theory, it was proved in 7.2 and 11.1 for the path-components X_k of X and $A_k = A \cap X_k$ that $\sum_k^{\oplus} H_q(X_k, A_k) = H_q(X, A)$. Cohomology has an analogous formula. In the axiomatic theory, only a weaker theorem can be proved.

12.6. THEOREM: *Let $X = \bigcup_{k=1}^n X_k$ be a finite decomposition of X into disjoint, open and closed subspaces X_k. Let $A \in X$ and $A \cap X_k = A_k$. The resulting space pairs and functions are admissible. Let $i_k\colon(X_k, A_k) \longrightarrow (X, A)$ be the injections. Then the $(i_k)_*$ yield an injective representation of $H_q(X, A)$ as a direct sum of the $H_q(X_k, A_k)$ and the i_k^* yield a projective representation of $H^q(X, A)$ as a direct product of the $H^q(X_k, A_k)$. In particular,*

$$H_q(X, A) \cong \sum\nolimits^{\oplus} H_q(X_k, A_k)$$

and

$$H^q(X, A) \cong \sum\nolimits^{\oplus} H^q(X_k, A_k).$$

Proof: (Incorrect for $n = 2$ in Eilenberg-Steenrod).

(1) Let $n = 2$; that is, $X = X_1 \cap X_2$ and $X_1 \cap X_2 = \varnothing$ where X_1 and X_2 are both open (closed) in X. For $A \subset X$ let $A_\nu = A \cap X_\nu$, $\nu = 1, 2$. Let $\mu = 3 - \nu$. Consider the (admissible) injections

$$h_\nu\colon(X_\nu, A_\nu) \longrightarrow (X_\nu \cup A_\mu, A);\ k_\nu\colon (X_\nu \cup A_\mu, A) \longrightarrow (X, X_\mu \cup A).$$

Here, h_ν is an excision in which $A_\mu = X_\mu \cap (X_\nu \cup A_\mu)$ is cut away. The formula shows that A_μ is simultaneously open and closed in the subspace $X_\nu \cup A_\mu$. The function $k_\nu h_\nu\colon(X_\nu, A_\nu) \longrightarrow (X, X_\mu \cup A)$ is also an excision, since it cuts away the open and closed subset X_μ from X. Consequently, $h_{\nu*}$ and $(k_\nu h_\nu)_*$ are isomorphisms; hence, $h_{\nu*}$ and $k_{\nu*}$ are isomorphisms. (The same can be said for h_ν^* and k_ν^*.) Now consider the diagram (in which for cohomology the arrows are to be reversed and the symbols q and $*$ raised):

$$\begin{array}{ccccc}
H_q(X, X_2 \cup A_1) & \xleftarrow{j_{2*}} & & \xrightarrow{j_{1*}} & H_q(X, X_1 \cup A_2) \\
\uparrow k_{1*} & & H_q(X, A) & & \uparrow k_{2*} \\
H_q(X_1 \cup A_2, A) & \xrightarrow{i'_{1*}} & & \xleftarrow{i'_{2*}} & H_q(X_2 \cup A_1, A) \\
\uparrow h_{1*} & & & & \uparrow h_{2*} \\
H_q(X_1, A_1) & & & & H_q(X_2, A_2)
\end{array}$$

Here i' and j are the injections and $i'_{\nu *} h_{\nu *} = i_{\nu *}$. The diagonals arise from the exact sequences of the space triple $(X, X_\mu \cap A_\nu, A)$. Lemmas 10.4 and 10.5 can be applied since the $k_{\nu *}$ are isomorphisms and the triangles are commutative (all functions are injections). Therefore, the $i'_{\nu *}$ yield an injective representation. Since the $h_{\nu *}$ are isomorphisms, the $i_{\nu *}$ also yield an injective representation, and

$$H_q(X, A) = i_{1*} H_q(X_1, A_1) \oplus i_{2*} H_q(X_2, A_2).$$

(2) Let $n > 2$, and assume the theorem proved for $q = n - 1$. Let $X' = \bigcup_{k=2}^n X_k$, $A' = \bigcup_{k=2}^n A \cap X'$, and let $i' : (X', A') \longrightarrow (X, A)$, $i'_k : (X_k, A_k) \longrightarrow (X', A')$, $k \geq 2$, be the injections. Then $i_k = i' i'_k$ for $k \geq 2$. It has to be proved that $a \in H_q(X, A)$ has only one representation $a = \sum i_{k*}(a_k)$ with $a_k \in H_q(X_k, A_k)$. From $\sum_{k=1}^n i_{k*}(a_k) = 0$, it follows that $i_{1*}(a_1) + i'_* \sum_{k=2}^n i'_{k*}(a_k) = 0$. By case (1), i_{1*} and i'_* form an injective representation of $H_q(X, A)$. Hence, $a_1 = 0$ and $\sum_{k=2}^n i'_{k*}(a_k) = 0$. By the inductive hypothesis, the i'_{k*}, $k \geq 2$ form an injective representation of $H_q(X', A')$. Therefore, $a_k = 0$ for $k \geq 2$. The proof for cohomology is analogous to the one for homology.

Special Case: Let X be a hausdorff space consisting of a finite set of points, say n, with the discrete topology. Theorem 12.6 yields $H_q(X) = 0$ and $H^q(X) = 0$ for $q \neq 0$. Furthermore, $H_0(X)$ and $H^0(X)$ are isomorphic to a direct sum of n summands M where $M = H_0(P)$ is the coefficient module. Let P be one of the points of X. For the reduced group (in other words, the augmented theory) the reduced homology sequence yields $\tilde{H}_0(X) \cong H_0(X, P)$. It was proved earlier that $H_0(P, P) = 0$, and Theorem 12.6 shows that $H_0(X, P)$, or $\tilde{H}_0(X)$, is isomorphic to a direct sum of $n - 1$ summands M. The same holds for $\tilde{H}^0(X)$.

Those familiar results of the singular theory which were founded upon homotopy theory also hold in the axiomatic theory; for, in the study of the purely topological concepts such as homotopic space pairs, deformation retracts, etc., only the homotopy axiom was needed. This shows that *homotopic space pairs have isomorphic homology groups and cohomology groups*. In particular, *a space which has a single point as deformation retract has the same groups as a point*. The computations of the groups of S^n and of the graphs carry over to the axiomatic theory with the same results, provided that the functions employed were admissible. The excision theorem was applied only in the cutting away of open sets in S^n and its consequences are therefore valid.

13

Mayer-Vietoris Sequence

Preliminary Considerations with Respect to Modules

13.1. THE HEXAGON THEOREM: *Let all triangles be commutative in the diagram of R-modules and R-homomorphisms at the top of the next page. Let the two slanting diagonals be exact, and let k_1 and k_2 be isomorphisms. Then $h_1k_1^{-1}l_1 + h_2k_2^{-1}l_2 = m_1m_0$.*

Proof: Lemmas 10.4 and 10.5 show that $i_1k_1^{-1}j_1 + i_2k_2^{-1}j_2 = \mathrm{Id}$. If this is applied to $m_0(c)$ for $c \in C$, and the relation $j_\nu m_0 = l_\nu$ is used, it follows that

$$m_1m_0(c) = m_1(i_1k_1^{-1}l_1(c) + i_2k_2^{-1}l_2(c)) = (h_1k_1^{-1}l_1 + h_2k_2^{-1}l_2)(c).$$

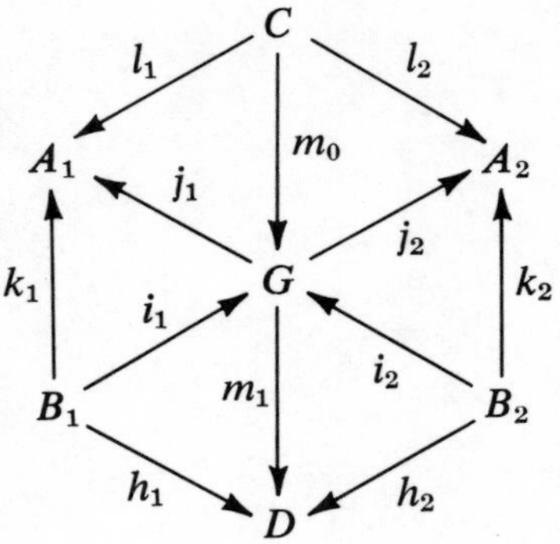

13.2. Corollary: *If* $m_1 m_0 = 0$, *then* $h_1 k_1^{-1} l_1 = -h_2 k_2^{-1} l_2$.

Theorem 13.1 will now be applied to Diagram 13.3:

13.3.

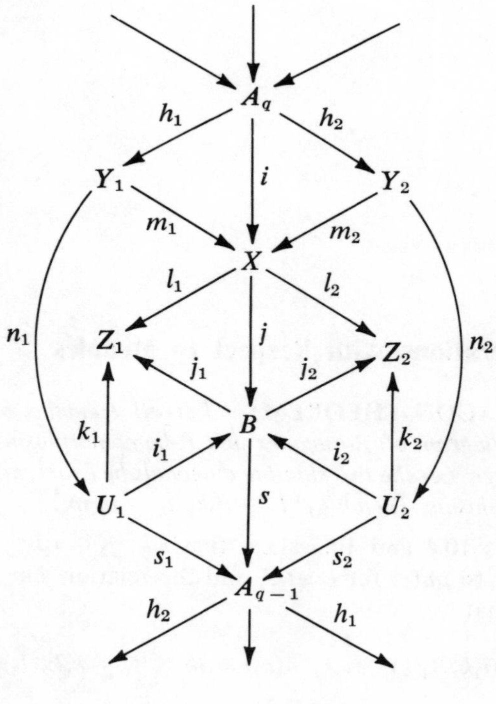

In this diagram, all of the modules and homomorphisms are to have indices q (these have been indicated at the A's only). All of the triangles are to be commutative. All indicated sequences of at least three modules, including the two outer contours, are to be exact. Furthermore, k_1 and k_2 are to be isomorphisms, and $i_1n_1 = jm_1$ and $i_2n_2 = jm_2$.

Additional homomorphisms

$$\psi:A_q \longrightarrow Y_1 \oplus Y_2, \quad \phi:Y_1 \oplus Y_2 \longrightarrow X, \text{ and } \Delta:X \longrightarrow A_{q-1}$$

are defined by $\psi(a) = (h_1a, -h_2a)$, $\phi(y_1, y_2) = m_1y_1 + m_2y_2$ and $\Delta(x) = -s_1k^{-1}l_1(x)$ where $a \in A_q$, $y_1 \in Y_1$, $y_2 \in Y_2$, and $x \in X$. The Hexagon Theorem (13.1) yields

13.4. THEOREM: *The sequence*

$$\cdots \xrightarrow{\Delta} A_q \xrightarrow{\psi} Y_1 \oplus Y_2 \xrightarrow{\phi} X \xrightarrow{\Delta} A_{q-1} \longrightarrow \cdots$$

is exact.

Proof:

(1) $\phi\psi(a) = m_1h_1(a) - m_2h_2(a) = i(a) - i(a) = 0$.

(2) $\Delta\phi(y_1, y_2) = s_2k_2^{-1}l_2m_1(y_1) - s_1k_1^{-1}m_2(y_2) = 0$, since $l_2m_1 = 0$ and $l_1m_2 = 0$.

(3) $\psi\Delta(x) = (-h_1s_1k_1^{-1}l_1(x), -h_2s_2k_2^{-1}l_2(x)) = 0$, since $h_1s_1 = 0$ and $h_2s_2 = 0$.

(4) Let $x \in X$ and let $\Delta(x) = 0$. Then $s_\nu(k_\nu^{-1}l_\nu(x)) = 0$ for $\nu = 1, 2$. Exactness of the outer contours shows the existence of a $y_\nu \in Y_\nu$ such that $n_\nu(y) = k^{-1}l_\nu(x)$. For $jx \in B$, Lemmas 10.4 and 10.5 yield

$$\begin{aligned} jx &= (i_1k_1^{-1}j_1 + i_2k_2^{-1}j_2)\, jx = i_1k_1^{-1}l_1x + i_2k_2^{-1}l_2x \\ &= i_1n_1y + i_2n_2y = jm_1y_1 + jm_2y_2. \end{aligned}$$

This shows that $j(x - m_1y_1 - m_2y_2) = 0$. By exactness, there is an $a \in A_q$ such that $ia = x - m_1y_1 - m_2y_2$. Therefore,

$$\begin{aligned} x &= ia + m_1y_1 + m_2y_2 \\ &= m_1(h_1a + y_1) + m_2y_2 = \phi(h_1(a) + y_1, y_2). \end{aligned}$$

(5) Let $\phi(y_1, y_2) = m_1y_1 + m_2y_2 = 0$. An $a \in A_q$ must be found for which $\psi(a) = (y_1, y_2)$. For this purpose, apply j and use

$jm = i_\nu n_\nu$. Then, $i_1n_1y_1 + i_2n_2y_2 = 0$. Since i_1 and i_2 form an injective representation of B as a direct sum, it follows that $n_1y_1 = 0$ and $n_2y_2 = 0$. By exactness, there are $a_\nu \in Y_\nu$ such that $h_\nu a_\nu = y_\nu$, $\nu = 1, 2$. The original equation yields $m_1h_1a_1 + m_2h_2a_2 = 0$, and therefore $i(a_1 + a_2) = 0$. By exactness, there is a b such that $a_1 + a_2 = s(b)$. Let b be written $b = i_1u_1 + i_2u_2$ where $u_\nu \in U_\nu$. Then,

$$a_1 + a_2 = si_1u_1 + si_2u_1 = s_1u_1 + s_2u_2.$$

The desired a is now

$$a = a_1 - s_1u_1 = -(a_2 - s_2u_2).$$

This can be seen from

$$\psi(a) = (h_1a_1, -h_2a_2) = (h_1a_1 - h_1s_1u_1, h_2a_2 - h_2s_2u_2) = (y_1; y_2)$$

which holds, since $h_\nu s_\nu = 0$ and $h_\nu a_\nu = y$ for $\nu = 1, 2$.

(6) If a satisfies $\psi(a) = (h_1a, h_2a) = 0$, an x must be found for which $\Delta(x) = 0$. When $\nu = 1, 2$, $h_\nu(a) = 0$. Exactness shows the existence of u_ν such that $a = s_\nu u_\nu$. From $si_\nu = s_\nu$, it follows that

$$s(-i_1u_1 + i_2u_2) = -s_1u_1 + s_2u_2 = -a + a = 0.$$

By exactness, there is an x for which $jx = -i_1u_1 + i_2u_2$. Then,

$$\begin{aligned}\Delta(x) = -s_1k_1^{-1}l_1(x) = -s_1k_1^{-1}j_1jx &= -s_1k_1^{-1}j_1(-i_1u_1 + i_2u_2)\\ &= s_1k_1^{-1}j_1i_1(u_1) = s_1(u_1) = a,\end{aligned}$$

because $j_1i_2 = 0$ and $j_1i_1 = k_1$.

Definition: A *triad* (X, X_1, X_2) consists of a topological space X and an ordered pair of subspaces X_1, and X_2, of X. The triad (X, X_1, X_2) is to be distinct from (X, X_2, X_1), provided that $X_1 \neq X_2$. The previously considered space triples were triads (X, A, B) for which $A \supset B$.

Each triad (X, X_1, X_2) has injections

$$k_1:(X_2, X_1 \cap X_2) \longrightarrow (X_1 \cup X_2, X_1)$$

and

$$k_2:(X_1, X_1 \cap X_2) \longrightarrow (X_1 \cup X_2, X_2).$$

The injection k_ν cuts away $X_\nu - (X_1 \cap X_2)$, $\nu = 1, 2$. The triad (X, X_1, X_2) is called *exact*, or a *proper triad*, if k_1 and k_2 are excisions; that is, if k_{1*} and k_{2*} (for cohomology, k_1^* and k_2^*) are isomorphisms in the non-augmented case. It is to be noticed that the concept of a proper triad is a function of the homology theory employed.

Remarks:

(1) In checking the exactness of a triad (X, X_1, X_2), the space X may be replaced by $X_1 \cup X_2$ since X does not occur in the definition of exactness.

(2) To an exact triad (X, X_1, X_2), there correspond isomorphisms, for instance $H_q(X_2, X_1 \cap X_2) \cong H_q(X_1 \cup X_2, X_1)$, which are clearly analogous to the isomorphism theorems of group theory.

Definition: The complement of a set A in a space X is denoted by $\mathscr{C}A$.

Examples:

(a) In R^{n+1} let $X = S^n$, $X_1 = E^n_+$ and $X_2 = E^n_-$. Then $X = X_1 \cup X_2$ and $S^{n-1} = X_1 \cap X_2$. Suppose that these spaces and the injection functions $k_1:(E^n_-, S^{n-1}) \longrightarrow (S^n, E^n_+)$ and $k_2:(E^n_+, S^{n-1}) \longrightarrow (S^n, E^n_-)$ are admissible. The considerations of 7.11 also hold in the axiomatic theory, and show that k_1 and k_2 are excisions. Consequently, (S^n, E^n_+, E^n_-) is an exact triad in the axiomatic theory.

(b) Let V_1 and V_2 be open subsets of a topological space X. In the singular theory, (X, V_1, V_2) is an exact triad.

Proof: Let $X = V_1 \cup V_2$. Let $U = \mathscr{C}V_2$ (the complement of V_2). Then $U = V_1 - (V_1 \cap V_2) \subset V_1$. Since U is closed and V_1 is open, the excision theorem of the singular theory shows that $k_1:(V_2, V_1 \cap V_2) \longrightarrow (V_1 \cup V_2, V_1)$ is an excision. For k_2, interchange the roles of V_1 and V_2.

(c) Let B_1, B_2 be compact subspaces of a hausdorff space X. Then B_1 and B_2 are closed. By (b), $(X, \mathscr{C}B_1, \mathscr{C}B_2)$ is an exact triad in the singular theory.

The theorem for Diagram 13.3 will now be applied to homology and cohomology groups. Let (X, X_1, X_2) be an exact triad in which $X = X_1 \cup X_2$. Let $A = X_1 \cap X_2$. In the case of homology, consider

13.5

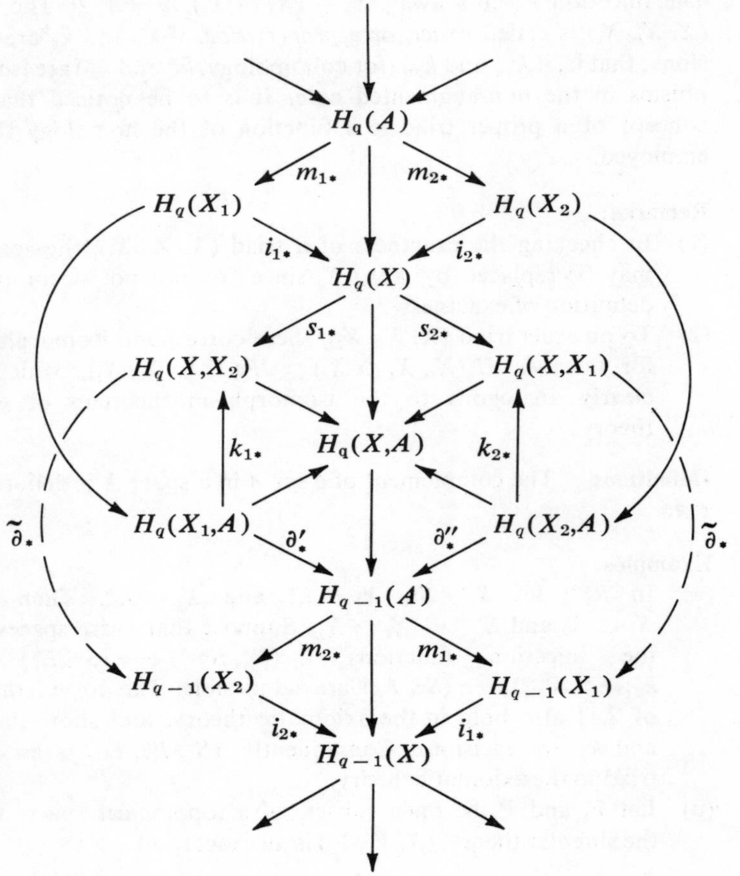

Diagram 13.5 which is the same as Diagram 13.3, except for the names of the modules and the two broken arrows. The indicated homomorphisms are boundary homomorphisms or else induced by injections and are assumed admissible. Since the triad (X, X_1, X_2) is exact, k_{1*} and k_{2*} are isomorphisms. The other conditions from Diagram 13.3 follow immediately from Axioms I, II, III, IV. Consequently, Theorem 13.2 can be applied in the case of homology.

The broken arrow $\tilde{\partial}_*: H_q(X, X_1) \longrightarrow H_{q-1}(X_1)$ yields a sequence

$$H_q(X) \longrightarrow H_q(X, X_1) \xrightarrow{\tilde{\partial}_*} H_{q-1}(X_1) \longrightarrow H_{q-1}(X)$$

which is exact by Axiom IV. By Axiom III, the rectangle

$$\begin{array}{ccc} H_q(X, A) & \longrightarrow & H_q(X, X_1) \\ \partial_* \downarrow & & \downarrow \partial_* \\ H_{q-1}(A) & \longrightarrow & H_{q-1}(X_1) \end{array}$$

is commutative.

In the case of cohomology, all of the arrows in 13.5 are to be reversed. If the diagram is turned upside-down and the old outer contours are replaced by the broken ones, then a diagram is obtained which coincides with 13.5. Hence, the hypotheses have been checked in the case of cohomology.

This has proved the following theorem:

13.6. MAYER-VIETORIS THEOREM:

(a) *Homology: Let* (X, X_1, X_2) *be an exact triad in which* $X = X_1 \cup X_2$, *and* $A = X_1 \cap X_2$. *Then the Mayer-Vietoris sequence*

$$\cdots \longrightarrow H_{q+1}(X) \xrightarrow{\Delta} H_q(A) \xrightarrow{\psi}$$
$$H_q(X_1) \oplus H_q(X_2) \xrightarrow{\phi} H_q(X) \xrightarrow{\Delta} H_{q-1}(A) \longrightarrow \cdots$$

is exact. Here ψ, ϕ, *and* Δ *are defined by* $\psi(a) = (m_{1*}a, -m_{2*}a)$, $\phi(x_1, x_2) = i_{1*}x_1 + i_{2*}x_2$, *and* $\Delta = -\partial'_* k_{1*}^{1-} s_{1*} = \partial'' k_{2*}^{-1} s_{2*}$.

(b) *Cohomology: Let* (X, X_1, X_2) *be an exact triad in which* $X = X_1 \cup X_2$, *and* $A = X_1 \cap X_2$. *Then the Mayer-Vietoris sequence*

$$\cdots \longleftarrow H^{q+1}(X) \xleftarrow{\Delta} H^q(A) \xleftarrow{\phi}$$
$$H^q(X_1) \oplus H^q(X_2) \xleftarrow{\psi} H^q(X) \xleftarrow{\Delta} H^{q-1}(A) \longleftarrow \cdots$$

is exact. Here ψ, ϕ, *and* Δ *are defined by* $\psi(x) = (i_1^*(x), -i_2^*(x))$, $\phi(x_1, x_2) = m_1^*(x_1) + m_2^*(x_2)$, *and* $\Delta = -s_2^* k_2^{*-1} \delta''^* = s_1^* k_1^{*-1} \delta'^*$.

In both cases all injections are assumed to be admissible.

Remark: If $X_1 \cap X_2 = A \neq \varnothing$, computations can also be made in the augmented theory. The reason is that only the property of being a monomorphism was used for k_{1*} and k_{2*}. The difference between augmented and non-augmented groups $H_q(Y, B)$ occurs only if $B = \varnothing$ and $q = 0$.

It should be noted that interchanging X_1 and X_2 leaves ϕ as it was, but changes the signs of ψ and Δ.

13.7. Corollary on Commutativity: *Let (X, X_1, X_2), and (X', X_1', X_2') be exact triads, in which $X = X_1 \cup X_2$ and $X' = X_1' \cup X_2'$. Let $f:(X, X_1, X_2) \longrightarrow (X', X_1', X_2')$ be admissible. With the homomorphisms induced by f, each rectangle in*

$$\begin{array}{ccccccccc}
\longrightarrow H_q(A) & \xrightarrow{\psi} & H_q(X_1) \oplus H_q(X_2) & \xrightarrow{\phi} & H_q(X) & \xrightarrow{\Delta} & H_{q-1}(A) \longrightarrow \\
\downarrow & & \downarrow \qquad\quad \downarrow & & \downarrow & & \downarrow \\
\longrightarrow H_q(A') & \xrightarrow{\psi} & H_q(X_1') \oplus H_q(X_2') & \xrightarrow{\phi} & H_q(X') & \xrightarrow{\Delta} & H_{q-1}(A') \longrightarrow
\end{array}$$

is commutative, and the analogous result holds for cohomology.

Proof: For all groups which occur in the Mayer-Vietoris Diagram 13.5, f defines homomorphisms which commute with the functions of that diagram (injections and boundary operators). Since ψ, ϕ, and Δ are expressible in terms of the functions of Diagram 13.5, the corollary follows.

13.8. Corollary (to the Mayer-Vietoris Theorem): *Let (X, X_1, X_2) be an exact triad in which $X = X_1 \cup X_2$. For some q, let $H_{q+1}(X) = 0$. Let $m_\nu : A = X_1 \cup X_2 \longrightarrow X_\nu$ be the injection, $\nu = 1, 2$. Let $a \in H_q(A)$ and $m_{1*}(a) = 0$, $m_{2*}(A) = 0$. Then $a = 0$.*

Proof: Since $\psi_q(a) = 0$, then $a \in \operatorname{Im} \Delta_{q+1}$. This is zero since $H_q(X) = 0$. Hence, $a = 0$.

Corollary 13.8 will be used in the proof of the Jordan-Brouwer Theorem in Chapter 14.

Application to Spheres: It has been shown that (S^n, E_+^n, E_-^n) is an exact triad in the axiomatic theory when $n \geq 0$. Consider the function $f:(S^n, E_+^n, E_-^n) \longrightarrow (S^n, E_+^n, E_-^n)$, which is given by

$$f(x_0, \cdots, x_{n-1}, x_n) = (x_0, \cdots, x_{n-1}, -x_n).$$

On $A = E_+^n \cap E_-^n = S^{n-1}$, f is the identity for $n \geq 0$. When $n \geq 1$, $A \neq \varnothing$. Therefore, the augmented theory can be applied, and $H_q(E_+^n) = H_q(E_-^n) = 0$. The Mayer-Vietoris sequence (13.7) now reduces, in the cases $n \geq 1$, to

$$\begin{array}{ccccccc}
0 \longrightarrow & H_q(S^n) & \xrightarrow{\Delta} & H_{q-1}(S^{n-1}) & \longrightarrow 0 \\
& \downarrow f_* & & \downarrow \mathrm{Id} & \\
0 \longrightarrow & H_q(S^n) & \xrightarrow{-\Delta} & H_{q-1}(S^{n-1}) & \longrightarrow 0.
\end{array}$$

The exactness shows that Δ is an isomorphism for all q. Hence, the groups $H_q(S^n)$ can be computed and the same results are obtained as in the singular theory (7.12). The groups $H^q(S^n)$ can be calculated in the analogous way.

The commutativity shows that $-\Delta f_* = \Delta$. Hence, $f_* = -\mathrm{Id}$ is a multiplication by (-1).

That $f_* = -\mathrm{Id}$, even when $n = 0$, can be seen as follows: Replace f in the case $n = 1$ by $g:(S^1, E^1_+, E^1_-) \longrightarrow (S^1, E^1_+, E^1_-)$ defined by $g(x_0, x_1) = (-x_0, x_1)$. The map g induces the map $f(x_0) = -x_0$ on S^0. Just as before in the case of f, the function g now induces a g_* such that $g_* = -\mathrm{Id}$. Now 13.7 yields the commutative, exact diagram

$$\begin{array}{ccccccc}
0 \longrightarrow & H_q(S^1) & \xrightarrow{\Delta} & H_{q-1}(S^0) & \longrightarrow 0 \\
& \downarrow g_* & & \downarrow f_* & \\
0 \longrightarrow & H_q(S^1) & \xrightarrow{\Delta} & H_{q-1}(S^0) & \longrightarrow 0.
\end{array}$$

Therefore, $f_*\Delta = \Delta g_* = -\Delta$ and $f_* = -\mathrm{Id}$. That $f^* = -\mathrm{Id}$ is proved analogously.

Application to Graphs: A graph is the union of a finite set of topological line segments which have at most end points in common. By the Direct Sum Theorem, it is sufficient to restrict considerations to connected graphs. Such a graph is homotopic to a graph G_r which consists of r circles attached at a point P since the remarks on graphs of Chapter 5 also hold in the axiomatic theory.

For $r \geq 2$, $G_r \supset G_{r-1}$ and $G_r \supset S^1$ in such a manner that

$$G_{r-1} \cup S^1 = G_r \text{ and } S^1 \cap G_{r-1} = \{P\}.$$

The reader can show as an exercise that (G_r, G_{r-1}, S^1) is an exact triad. The Mayer-Vietoris sequence for augmented complexes

$$0 = H_q(P) \xrightarrow{\psi} H_q(S^1) \oplus H_q(G_{r-1}) \xrightarrow{\phi} H_q(G_r) \xrightarrow{\Delta} H_{q-1}(P) = 0$$

shows that $H_q(G_r) \cong H_q(S^1) \oplus H_q(G_{r-1})$ for all q. Induction on r yields

$H_q(G_r) = 0$ for $q \neq 1$ (in the augmented case)

$H_1(G_r) \cong \sum^{\oplus} M$, with r summands M.

The result is the same for cohomology.

Application to Surfaces of Genus g:

Definition: Let X_1 and X_2 be topological spaces and let U_1 and U_2 be homeomorphic n-dimensional euclidean neighborhoods for which $U_1 \subset X_1$ and $U_2 \subset X_2$. Let E_1^n be a ball in U_1 and let E_2^n be its homeomorphic image in U_2. Let X be the space obtained by attaching $X_1 - \mathring{E}_1^n$ to $X_2 - \mathring{E}_2^n$ by means of the homeomorphism between E_1^n and E_2^n. The space X is said to be obtained by *adjoining* X_1 and X_2. If X and a third space X_3 are adjoined by the same procedure, the resulting space is said to have been obtained by the *successive adjunction* of the spaces X_1, X_2, and X_3.

Definition: Let F_0 be homeomorphic to S^2 and let F_g arise from F_{g-1} (up to a homeomorphism) by means of adjoining a torus. Each homeomorphic image in R^3 of such an F_g is called a surface of *genus* g. It is referred to as a sphere with g "handles." (In R^3 the handles can be knotted and intertwined.)

Let F_g be a surface of genus g. That such a surface always separates R^3 into two disjoint open sets is not always obvious. However, it is obvious in case F_g is represented in R^3 as in Figure 15. Let J be the closure of the bounded complementary domain of F_g and let J' be the closure of the unbounded complementary domain. J and J' are referred to as the "inside" and "outside" of F_g, respectively. Let R^3 be compactified to a homeomorph of S^3 by the addition of a single point. This compactifies J', whose designation will remain the

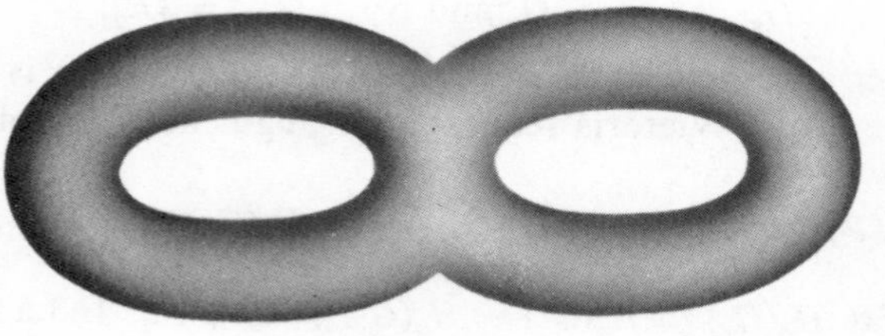

Figure 15

same. Then $J \cup J' = S^3$ and $A = J \cap J' = F_g$. The reader should show as an exercise that (S^3, J, J') is an exact triad, that J is homotopic to a graph G_g, and that J' is also homotopic to a graph G_g. Therefore, $H_q(J) \oplus H_q(J') = 0$ for $q \neq 1$, and $H_1(J) \oplus H_1(J')$ is isomorphic to a direct sum of $2g$ summands M (in the augmented case). Since $A \neq \varnothing$, the Mayer-Vietoris sequence can be employed for the augmented case (reduced homology groups), and it follows that

13.9.
$$H_q(F_q) = 0 \text{ for } q \neq 1, 2 \text{ (in the augmented case).}$$
$$H_2(F_g) \cong M.$$
$$H_1(F_g) \cong \text{direct sum of } 2g \text{ summands } M.$$

For cohomology there is an identical result since the direction of the arrows does not matter here.

13.10. Corollary: *If M is chosen to be a field, it follows that surfaces of different genus are not homeomorphic or even homotopic.*

Remark: In the (augmented) computations just carried out, the symbol $\tilde{H}_0(F_g)$ should actually have been used. It differs from the non-reduced group $H_0(F_g)$ by only one direct summand.

13.11. The Relative Mayer-Vietoris Sequence: *Let (X, X_1, X_2) be an exact triad, where it is not necessary that $X = X_1 \cup X_2$. Let $Y = X_1 \cup X_2$ and let $A = X_1 \cap X_2$. Then the relative Mayer-Vietoris sequence in homology*

$$\xrightarrow{\Delta} H_q(X, A) \xrightarrow{\psi} H_q(X, X_1) \oplus H_q(X, X_2) \xrightarrow{\phi} H_q(X, Y) \xrightarrow{\Delta} H^{q-1}(X, A) \longrightarrow$$

and the relative Mayer-Vietoris sequence in cohomology

$$\xleftarrow{\Delta} H^q(X, A) \xleftarrow{\phi} H^q(X, X_1) \oplus H^q(X, X_2) \xleftarrow{\psi} H^q(X, Y) \xleftarrow{\Delta} H^{q-1}(X, A) \longleftarrow$$

are exact.

Proof: Consider the diagram which arises from 13.5 when the groups of the sequence $H_q(A)$, $H_q(X_1)$, $H_q(X_2)$, $H_q(X)$, $H_q(X, X_2)$,

$H_q(X, X_1)$, $H_q(X, A)$ are respectively replaced by the groups of the sequence $H_q(X, A)$, $H_q(X, X_1)$, $H_q(X, X_2)$, $H_q(X, Y)$, $H_q(Y, X_2)$, $H_q(Y, X_1)$, $H_q(Y, A)$, in that order. The functions here are either the boundary operators or else are induced by the injections (which are assumed admissible). All the necessary hypotheses are satisfied by the new diagram and therefore the theorem is proved.

As in 13.7, admissible functions $f:(X, X_1, X_2) \longrightarrow (X', X_1', X_2')$ induce functions f_* on the exact sequences for which the rectangles in the diagram are commutative.

14

The Jordan-Brouwer Separation Theorems

This chapter makes use of the singular theory. The triads (X, X_1, X_2) considered consist of the complements of compact subsets of S^n. Part (c) of the example preceeding 13.5 establishes the exactness of such triads. In order that the augmented theory be applicable, considerations are restricted to the case $X_1 \cap X_2 \neq \varnothing$.

Let X, Y be spaces with $X \subset Y$ and let $i:X \longrightarrow Y$ be the injection. If c is a q-chain in X; that is, a linear combination of simplices $\sigma:\Delta_q \longrightarrow X$ with coefficients in M, then $i^{\#}c$ is a linear combination of simplices which arise from σ by an extension of the image space from X to Y. The notation will be simplified by the omission of $i^{\#}$, and c will be called a chain in X or in Y depending upon the situation.

The closed ball $E^{r+1} = \{x \mid x_0^2 + \cdots + x_r^2 \leq 1\}$ is homeomorphic to a cube I^{r+1}, where I is the interval $0 \leq t \leq 1$. A homeomorphism of I^{r+1} on E^{r+1} is given by $\phi(x) = (x/|x|)\cdot \text{Max}_i\{|x_i|\}$ for $x \neq 0$ and $\phi(0) = 0$. I^0 is understood to be a point.

14.1. THEOREM: *For $n \geq 0$, let S^n be the n-sphere and let $\tilde{E}^r \subset S^n$ be a homeomorphic image of I^r, $r \geq 0$. Then, in the augmented singular theory,*

$$H_q(\mathscr{C}\tilde{E}^r) = 0 \text{ for all } q \text{ and } r.$$

Proof:

(1) The set $\mathscr{C}\tilde{E}^0$ is homotopic to a point. Therefore, $H_q(\mathscr{C}\tilde{E}^0) = 0$ in the augmented case. The proof now proceeds by induction on r.

(2) Let c be a q-cycle in $\mathscr{C}\tilde{E}^r$. For each $t \in I$, the image of $\{t\} \times I^{r-1}$ is an $\tilde{E}^{r-1}(t)$. Since $\mathscr{C}\tilde{E}^{r-1}(t) \supset \mathscr{C}\tilde{E}^r$, c can be viewed as a cycle in $\mathscr{C}\tilde{E}^{r-1}(t)$. By the inductive hypothesis, $H_q(\mathscr{C}^{r-1}(t)) = 0$; therefore c is a boundary. Hence, there is a $(q + 1)$-chain b_t in $\mathscr{C}\tilde{E}^{r-1}(t)$ for which $c = \partial b_t$. The carrier $|b_t|$ of b_t is a compact and therefore closed subset which does not meet $\tilde{E}^{r-1}(t)$. Denote the distance from $|b_t|$ to $\tilde{E}^{r-1}(t)$ in the metric space S^n by ϵ_t. Then $\epsilon_t > 0$. Each continuous function of the compact set I^r in S^n is uniformly continuous. Therefore, there is a $\delta(t) > 0$, such that points in I^r whose distance is less than $\delta(t)$ have images in S^n whose distance is less than ϵ_t. Now let $I' \subset I$ be an open interval of length less than $2\delta(t)$ and with midpoint t. Let $\tilde{E}^r(t)$ be the image of $I' \times I^{r-1}$. Each point of $\tilde{E}^r(t)$ has a distance less than ϵ_t from $\tilde{E}^{r-1}(t)$. Therefore, $|b_t|$ does not meet the set $\tilde{E}^r(t)$. Therefore, $c = \partial b_t$ in $\mathscr{C}\tilde{E}^r(t)$. The intervals I' cover I. Since I is compact there is a $\rho > 0$ such that each interval of length $< \rho$ lies in an I' (Theorem 6.11). Choose a natural number m with $(1/m) > \rho$. Let I_ν be the interval $(\nu/m) \leq t \leq (\nu + 1)/m$, $\nu = 0, \cdots, m - 1$. Let $\tilde{E}_\nu^r$ be the image of $I_\nu \times I^{r-1}$. Then there is a chain b_ν in $\mathscr{C}\tilde{E}_\nu^r$ for which $c = \partial b_\nu$.

At this point, it is necessary to digress to prove a lemma.

14.2. Lemma: *Let J_1 and J_2 be closed subintervals of I such that $J_1 \cap J_2 = \{t\}$. Let $\tilde{E}'$ be the image of $J_1 \times I^{r-1}$ and let $\tilde{E}''$ be the image of $J_2 \times I^{r-1}$. In $\mathscr{C}\tilde{E}'$ let b' be a $(q + 1)$-chain for which $c =$*

$\partial b'$. In $\mathscr{C}\tilde{E}''$ let b'' be a $(q+1)$-chain for which $c = \partial b''$. Then there is a $(q+1)$-chain b in $\mathscr{C}(\tilde{E}' \cap \tilde{E}'')$ for which $c = \partial b$.

Proof: Let $X = \mathscr{C}\tilde{E}^{r-1}(t)$, $X_1 = \mathscr{C}\tilde{E}'$, and $X_2 = \mathscr{C}\tilde{E}''$. Then,

$$A = X_1 \cap X_2 = \mathscr{C}(\tilde{E}' \cup \tilde{E}'').$$

Since $\tilde{E}' \cup \tilde{E}''$ and S^n have different homology groups ($\tilde{E}' \cup \tilde{E}''$ is homotopic to a point), it follows that

$$\mathscr{C}(\tilde{E}' \cup \tilde{E}'') = A \neq \varnothing.$$

When $X_1 \cup X_2 = \mathscr{C}(\tilde{E}' \cap \tilde{E}'')$ is called X, Example (c) shows that (X, X_1, X_2) is an exact triad. By inductive hypothesis,

$$H_{q+1}(X) = H_{q+1}(\mathscr{C}\tilde{E}_{r-1}(t)) = 0.$$

The cycle c was a boundary in X_1 and in X_2, and by Corollary 13.8, c is boundary of a chain b in $A = \mathscr{C}(\tilde{E}' \cup \tilde{E}'')$.

Now return to the proof of 14.1: Repeated application, $(m-1)$ times, of the lemma to the sets $\tilde{E}^r_\nu$ shows that c is the boundary of a chain b in $\mathscr{C}(\bigcup_{\nu=0}^{m-1}\tilde{E}^r_\nu) = \mathscr{C}\tilde{E}^r$. Since c was an arbitrary cycle in $\mathscr{C}\tilde{E}^r$ and had a boundary there, it follows that $H_q(\mathscr{C}\tilde{E}^r) = 0$ for all q. This completes the proof of 14.1.

Suppose that $\tilde{S}^r$, $1 \leq r \leq n$, is the homeomorphic image of the S^r in S^n. The images of E^r_+, E^r_-, and S^{r-1} will be denoted by $\tilde{E}^r_+$, $\tilde{E}^r_-$, and $\tilde{S}^{r-1}$ ($\tilde{S}^{r-1} = \tilde{E}^r_+ \cap \tilde{E}^r_-$). In the case $r = n$, it will also be assumed that $\tilde{S}^n \neq S^n$. Let $X = \mathscr{C}\tilde{S}^{r-1}$, $X_1 = \mathscr{C}\tilde{E}^r_+$ and $X_2 = \mathscr{C}\tilde{E}^r_-$. The triad (X, X_1, X_2) is exact in the singular theory. When $r < n$, the nth homology group of $\tilde{S}^r$ and S^n are different. Hence, $\tilde{S}^r \neq S^n$, and

$$A = X_1 \cap X_2 = \mathscr{C}\tilde{S}^r \neq \varnothing$$

for $1 \leq r \leq n$. Consequently, computations in the augmented theory are permissible. Since E^r_+ and E^r_- are homeomorphic to I^r, it follows from 14.1 that $H_q(X_1) = H_q(X_2) = 0$. From this and the Mayer-Vietoris sequence, it follows that $H_{q+1}(X) \cong H_q(A)$. Hence, $H_{q+1}(\mathscr{C}\tilde{S}^{r-1}) \cong H_q(\mathscr{C}\tilde{S}^r)$ for $1 \leq r < n$ and also for $r = n$ when $\tilde{S}^n \neq S^n$. Therefore, $H_q(\mathscr{C}\tilde{S}^r) \cong H_{q+r}(\mathscr{C}\tilde{S}^0)$. Since S^0 consists of two points which can be moved by a homotopy to the north and south poles, respectively, $\mathscr{C}\tilde{S}^0$ is homotopic to S^{n-1}. Therefore,

$$H_q(\mathscr{C}\tilde{S}^r) \cong H_{q+r}(S^{n-1}) \cong \begin{cases} M & \text{for } q = n - r - 1 \\ 0 & \text{otherwise.} \end{cases}$$

In the case $r = n$, this implies that $H_{-1}(\mathscr{C}\tilde{S}^n) \cong M$, a contradiction to $H_{-1} = 0$ (in the singular theory). Consequently, $\tilde{S}^n = S^n$. It also follows from this that S^n contains no homeomorphic image of S^r for $r > n$. For such an $\tilde{S}^r$ would have a proper subspace $\tilde{S}^n$ which would have to be equal to S^n in contradiction to the one-to-one property of homeomorphisms. These results are summarized in the next theorem.

14.3. THEOREM: *When $r > n$, S^n contains no homeomorphic image of S^r. When $r = n$, S^n contains only itself as a homemorphic image of S^n. When $0 \leq r < n$,*

$$H_q(\mathscr{C}\tilde{S}^r) \cong \begin{cases} M & \textit{for } q = n - r - 1 \\ 0 & \textit{otherwise.} \end{cases}$$

The augmented homology groups of $\mathscr{C}\tilde{S}^r$ are thus those of an $(n - r - 1)$-dimensional sphere.

14.4. Corollary (The Jordan-Brouwer Separation Theorem): *If $n \geq 1$, the complement of $\tilde{S}^{n-1}$ in S^n consists of exactly two pathwise-connected components, each of which has $\tilde{S}^{n-1}$ as boundary.*

Proof:

(1) Let M be a field Q. Then $H_0(\mathscr{C}\tilde{S}^{n-1}) = Q$ in the augmented theory. By 11.2, the number r of path-components is 2.

(2) Since $\tilde{S}^{n-1}$ is compact, $\mathscr{C}\tilde{S}^{n-1}$ and its path-components K_1 and K_2 are open. The boundary of each component is disjoint from the other component, and is therefore contained in $\tilde{S}^{n-1}$. Conversely, let $x \in \tilde{S}^n$. It will be proved that x is a boundary point of both K_1 and K_2; in other words, that each neighborhood U of x contains points of K_1 and of K_2. Let $y_1 \in K_1$ and $y_2 \in K_2$. Let $x_0 \in S^{n-1}$ be the inverse image of x. Choose an open disc (a cap in this case) $K \subset S^{n-1}$ about x_0, whose image $\tilde{K}$ is contained entirely in U. Then $\tilde{S}^{n-1} - \tilde{K} = \tilde{E}^{n-1}$ is homeomorphic to an I^{n-1}. From Theorem 14.1, it follows that $H_0(\mathscr{C}\tilde{E}^{n-1}) = 0$. Therefore, $\mathscr{C}\tilde{E}^{n-1}$ is pathwise-connected, and

there is a path Γ in $\mathscr{C}\tilde{E}^{n-1}$ connecting y_1 to y_2. Since y_1 and y_2 lie in distinct component of $\mathscr{C}\tilde{S}^{n-1}$ this path must meet $\tilde{S}^{n-1}$; and, indeed, since $\Gamma \cap \tilde{S}^{n-1} = \varnothing$, the path must pass through $\tilde{K}$. The set $\Gamma \cap \tilde{S}^{n-1} = \Gamma \cap \tilde{K}$ is closed and not empty. Hence, the inverse image of $\Gamma \cap \tilde{S}$ is a closed and therefore compact subset of I. Hence, there is a first intersection point s_1 and a last intersection point s_2 of Γ with $\tilde{K}$. In each neighborhood of $s_i \in \tilde{K} \subset U$, and therefore in U, there are points of K_i; for y_i is connected with s_i by a path which, except for the endpoint s_i, is completely contained in K_i, $i = 1, 2$. The following figure shows that the theorem is not trivial even in the case $n = 2$.

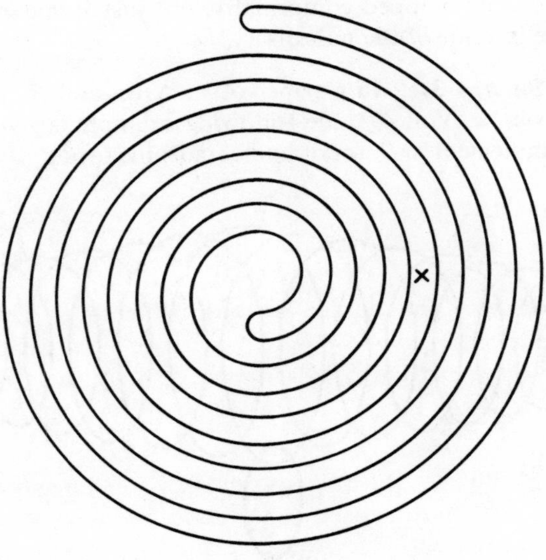

Figure 16

Remarks: By 7.2, the homology groups of the components of $\mathscr{C}\tilde{S}^{n-1}$ are zero for $q \geq 1$. The 0-dimensional group of each component is 0 in the augmented and isomorphic to M in the non-augmented case. Thus, both components have the same homology groups as in $\mathring{E}^{n-1}$. In the case $n = 2$, each component can be proved homeomorphic to $\mathring{E}^1$. However, there are counterexamples when $n \geq 3$.

14.5. THEOREM: *The components of the complement of an* $\tilde{S}^1$ *in* S^2 *are homeomorphic to open discs.*

Indication of proof: Let $\tilde{S}^1 \subset S^2$. Interpret S^2 as the complex plane in which the point at infinity is one of the points not on $\tilde{S}^1$. The bounded component A of the Jordan curve $\tilde{S}^1$ will be called the interior of $\tilde{S}^1$. Let $f(z)$ be a regular function that has no zeroes in A. By the Cauchy Integral Theorem (11.9) the function defined by $\sqrt{f(z)}$ can be uniquely determined by

$$\sqrt{f(z)} = \sqrt{f(z_0)} \exp \frac{1}{2} \int_{z_0}^{z} \frac{f'(z)}{f'(z)} dz,$$

where $\sqrt{f(z_0)}$ is arbitrarily chosen (but fixed). This is the essential step in the proof of the Riemann Mapping Theorem, according to which A can be mapped conformally, not just homeomorphically, on E^1, the interior of the unit disc.

Example for $n = 3$: In a paper of E. Artin and R. Fox (*Annals of Mathematics*, Vol. 49), the following example is given of an $\tilde{E}^1$ whose complement in S^3 is not homeomorphic to $\mathring{E}^3$.

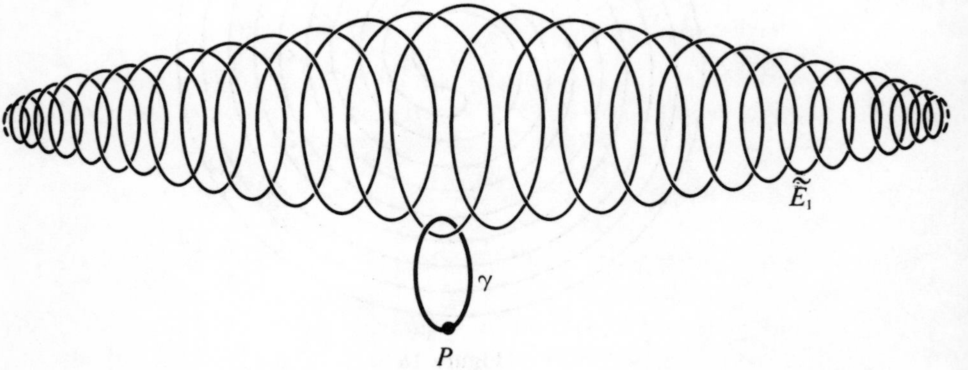

Figure 17

Here, $\tilde{E}^1$ is homeomorphic to a line segment. It is plausible, and can indeed be proved, that the curve γ which lies in $\mathscr{C}\tilde{E}^1$ can not be contracted in $\mathscr{C}\tilde{E}^1$ to the point P, which the point held fixed during the contraction. This would be possible in $\mathring{E}^3$.

From this example an $\tilde{S}^2 \subset S^3$ can be constructed whose outer component is not homeomorphic to E^3. For this purpose, replace the curve $\tilde{E}^1$ by a closed tube which gradually becomes more narrow towards the vertices. The inside of this $\tilde{S}^2$ is homeomorphic to E^3 but not the outside. Now make a small hole in the $\tilde{S}^2$ and attach a short tube which lies in the interior of $\tilde{S}^2$. To the free end of the tube attach a smaller copy of the $\tilde{S}^2$. In this way one obtains a homeomorphic image of $\tilde{S}^2$ for which neither the inside nor the outside is homeomorphic to E^3. There are analogous examples for $n > 3$.

A homeomorphic image $\tilde{S}^1$ of S^1 in S^3 is called a *knot*. It has been proved that $H_1(\mathscr{C}\tilde{S}^1) \cong M$, and $H_q(\mathscr{C}\tilde{S}^1) = 0$, for $q \neq 1$, in the augmented theory. The homology groups $H_q(\mathscr{C}\tilde{S}^1)$ are therefore as useless as the groups $H_q(\tilde{S}^1)$ in distinguishing among knots.

Definitions: An open, pathwise-connected (i.e. an open, connected) subset of R^n is called a *region*.

If $\tilde{S}^{n-1} \subset S^n$ and $n \geq 1$, then the Jordan-Brouwer Theorem states that $S^n - \tilde{S}^{n-1}$ consists of two disjoint regions. Pick a point x of one of these regions. Then $S^n - \{x\}$ is homeomorphic to R^n. The complementary region containing x will be referred to as the *outside* of $\tilde{S}^{n-1}$, and the other as the *inside*. Removing x from its region will disconnect that region when $n = 1$, but not otherwise, for a path joining any two points of a region can be modified slightly to avoid a given point. This yields the next corollary.

14.6. Corollary: *Let $n \geq 2$, and let $\tilde{S}^{n-1}$ be a homeomorphic image of S^{n-1} in R^n. Then $\tilde{S}^{n-1}$ consists of two regions, its path components, each of which has $\tilde{S}^{n-1}$ as boundary.*

14.7. Corollary: *Let $n \geq 2$ and let $f\colon E^n \longrightarrow R^n$ be a continuous, one-to-one function. Let $\tilde{S}^n = f(S^{n-1})$ and let A be the interior of the region determined by S^n. Then $f(\mathring{E}^n) = A$.*

Proof: (That this corollary is not trivial should be clear from the "tube" generated from the Fox-Artin example which was an $\tilde{S}^2 \subset S^3$ for which neither complementary domain was topologically an E^3).

Since S^n is a hausdorff space and E^n is compact, it follows that f is a homeomorphism. Let $\tilde{E}^n = f(E^n)$. Theorem 14.1 states that $H_q(\mathscr{C}\tilde{E}^n) = 0$ in the augmented theory. In particular, this means that $\mathscr{C}\tilde{E}^n = S^n - \tilde{E}^n$ is pathwise-connected. Now embed R^n in S^n (by removing a point of $\mathscr{C}\tilde{E}^n$ from S^n). Then, as in the proof of 14.6, $R^n - \tilde{E}^n$ remains pathwise-connected. The set $\tilde{E}^n$ is compact; hence, closed and bounded. Hence, $R^n - \tilde{E}^n$ is not bounded (contains a deleted neighborhood of the removed point). Since $\tilde{S}^{n-1} \subset \tilde{E}^n$, it follows that $(R^n - \tilde{E}^n) \cap \tilde{S}^{n-1} = \varnothing$ and that $(R^n - \tilde{E}^n) \subset B$, where B is the outside of $\tilde{S}^{n-1}$. Therefore, $\tilde{E}^n \supset R^n - B = \tilde{S}^{n-1} \cup A$. Since f is one-to-one, $f(\mathring{E}^n)$ and $\tilde{S}^{n-1}$ are disjoint and therefore, $f(\mathring{E}^n) \supset A$ where A is the inside of $\tilde{S}^{n-1}$. As the continuous image of a pathwise-connected set, $f(E^n)$ is pathwise-connected. It was already known to be disjoint from $\tilde{S}^{n-1}$ and therefore lies entirely in A or else entirely in B. Since $A \subset f(E^n)$, the only possibility is $f(E^n) = A$.

14.8. Corollary (Invariance of Domain): *Let G be a region in R^n and let $f: G \longrightarrow R^n$ be continuous and one-to-one. Then f is an open function, $f(G)$ is a region, and f is a homeomorphism of G onto $f(G)$.*

Proof. Let $U \subset G$ be an open set and $y = f(x) \in f(U)$. Let $E^n \subset U$ be a closed ball about x with positive radius. By 14.7, $f(\mathring{E}^n)$ is the inside of $f(\tilde{S}^{n-1})$ and therefore, an open set in R^n which contains y and lies in $f(U)$. Hence, $f(U)$ is open, f is open, f is a homeomorphism, and finally, as image of a pathwise-connected set, $f(G)$ is a region.

Remark: If $f: R^1 \longrightarrow R^2$ is continuous and one-to-one, the image of an open set does not need to open in $f(R^1)$.

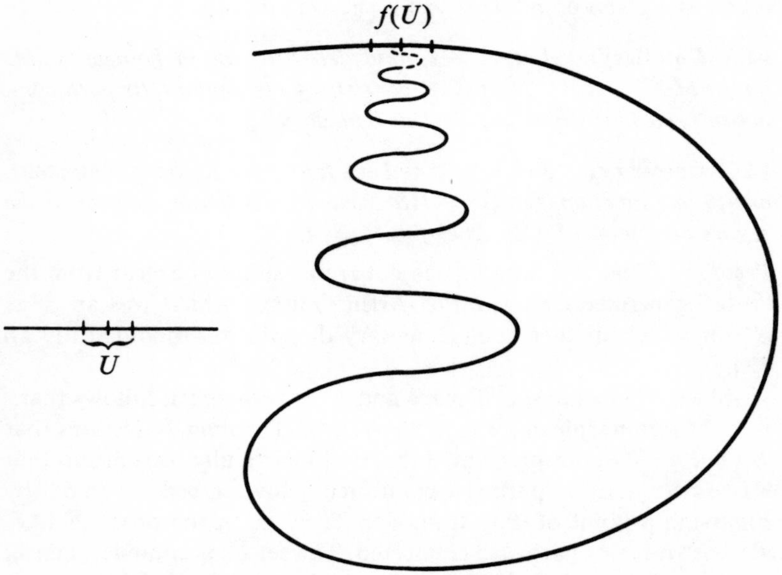

Figure 18

14.9. Corollary (Invariance of Dimension): *Let $m < n$, Then there is no continuous, one-to-one function f sending a region of R^n into R^m.*

Proof: Embed R^m in R^n. Then such an f would yield a function sending a region G in R^n into R^n. By 14.8, $f(G)$ is a region in R^n, but, by $f(G) \subset R^m \subset R^n$, it follows that $f(G)$ is not open.

15

Finite Cell Complexes

This chapter describes a process for building more complicated spaces from simple ones by the device of attaching n-cells. The homology theory employed will be the axiomatic one, and all spaces and functions that appear are to be admissible.

Let E^n be the unit ball in R^n and let S^{n-1} be its boundary. Let X_0 be a topological space for which $X_0 \cap E^n = \varnothing$, and let $f: S^{n-1} \longrightarrow X_0$ be continuous. Let $X = X_0 \cup \mathring{E}^n$. Let $i: X_0 \longrightarrow X$ be the injection. Now define $g: E^n \longrightarrow X$ by $g(x) = x$ for $x \in \mathring{E}^n$ and $g(x) = f(x)$ for $x \in S^{n-1}$. The functions i and f define a composite function $f_1: X_0 \cup E^n \longrightarrow X$. A set V is open in X if and only if $f_1^{-1}(V)$ is open in $X_0 \cup E^n$. This defines a topology in X, and f_1 is then a continuous function. The space X is said to have been obtained by *attaching the n-cell* E^n *to* X_0 *by means of* g. The topology defined on X is clearly the smallest topology for which f_1 is continuous.

If V is a set in X, then $V = (V \cap X_0) \cup (V \cap \mathring{E}^n)$.

15.1. *Denote* $V \cap X_0$ *by* V_0 *and* $V \cap \mathring{E}^n$ *by* V_1. *Then* V *is open in* X *if and only if*

(1) $i^{-1}(V) = V_0$ *is open in* X_0.

(2) $g^{-1}(V) = g^{-1}(V_0) \cup g^{-1}(V_1) = f^{-1}(V_0) \cup V_1$
$= f^{-1}(V_1 \cap f(S^{n-1})) \cup V_1$ *is an open set in* E^n.

In the special case that $V \subset \mathring{E}^n$, it is clear that $V_0 = \varnothing$, $V_1 = V$, and $g^{-1}(V) = V$. Hence, the topology defined on X induces the old metric topology on $\mathring{E}^n$ and $\mathring{E}^n$ is an open subspace of X.

As complement of $\mathring{E}^n$, X_0 is a closed subspace of X. It will now be proved that the topology induced on X_0 is also the original topology, or in other words, that X_0 *is a subspace of* X: If $V \subset X$ is open, then $V_0 = V \cap X_0 = i^{-1}(V)$ is open in X_0 by (1) of 15.1. On the other hand, if V_0 is an open subset of X_0, then $f^{-1}(V_0)$ is an open subset of S^{n-1} by the continuity of f. Hence, $f^{-1}(V_0) = W \cap S^{n-1}$ where W is open in E^n. Then $V_1 = W \cap \mathring{E}^n$ is open in E^n, and W is the disjoint union of V_1 and $f^{-1}(V_0)$. Now let $V = V_0 \cup V_1$. Here, V_0 is open in X_0 and $f^{-1}(V_0) \cup V_1 = W$ is open in E^n. Therefore, by 15.1, V is open in X. Since $V_1 \subset \mathring{E}^n$, it follows that $V_0 = V \cap X_0$. Hence, V_0 is open in the topology induced by X on X_0.

In the special case that $X_0 = S^{n-1}$ and $f = \text{Id}$, then $X = E^n$ and $g = \text{Id}$.

15.2. Lemma: *If* X_0 *is a hausdorff space, then* X *is a hausdorff space.*

Proof: Let x_1 and x_2 be points of X for which $x_1 \neq x_2$. There are three cases to be considered: (a) Both x_1 and x_2 are in $\mathring{E}^n$. Then they can be separated by neighborhoods contained in $\mathring{E}^n$. (b) They lie one in each part of X, say $x_1 \in X_0$ and $x_2 \in \mathring{E}^n$. Then choose t such that $|x_2| < t < 1$ and let $U = \{x \mid x \in \mathring{E}^n, |x| < t\}$. As a neighborhood of x_1, choose $X_0 \cup B_t$ where $B_t = \{x \mid x \in \mathring{E}^n, t < |x| < 1\}$. Since X_0 is open in X_0 and $f^{-1}(X_0) \cup B_t = S^{n-1} \cup B_t$ is open in E^n, it follows that $X_0 \cup B_t$ is open in X. (c) Both x_1 and x_2 lie in X_0. Then let V_1 and V_2 be disjoint open subsets of X_0 that separate x_1 and x_2. Since $V_1 \cap V_2 = \varnothing$, $f^{-1}(V_1) \cap f^{-1}(V_2) = \varnothing$. Define V_i' by $V_i' = \{x \mid x \in E^n, x/|x| \in f^{-1}(V_i), |x| > \frac{1}{2}\}$. Then $V_i \cup V_i'$, $i = 1, 2$, are open subsets of X which separate x_1 and x_2.

15.3. Lemma: *If X_0 is compact, then X is compact.*

Proof: The continuity of g and the compactness of E^n yield the compactness of $g(E^n)$. The continuity of i guarantees the compactness of X_0 as a subset of X. Hence, $X = X_0 \cup g(E^n)$ is compact.

Let the sets B, B', and B'' be defined by $B = \{x \mid \frac{1}{2} < |x| < 1\}$, $B' = \{x \mid \frac{3}{4} < |x| < 1\}$, and $B'' = \{x \mid \frac{3}{4} \leq |x| < 1\}$. They are subsets of $\mathring{E}^n$. By 15.1, $X_0 \cup B$ is open since X_0 is open in X_0 and $f^{-1}(X_0) \cup B = S^{n-1} \cup B$ is open in E^n. Similarly, $X \cup B'$ is open in X. On the other hand, $X_0 \cup B''$ is closed in X since its complement is open. Clearly, $X_0 \cup B' \subset X_0 \cup B'' \subset X_0 \cup B$. Therefore, the Excision Axiom applies and $X_0 \cup B'$ can be cut away; that is, the injection $(\mathring{E}^n - B', B - B') \longrightarrow (X, X_0 \cup B)$ is an excision and induces an isomorphism of the homology groups. In the special case that $X_0 = S^{n-1}$, the injection $(\mathring{E}^n - B', B - B') \longrightarrow (E^n, S^{n-1} \cup B)$ is also an excision.

The function g induces the following:

g':$(E^n, S^{n-1} \cup B) \longrightarrow (X, X_0 \cup B)$, g'':$(\mathring{E}^n - B', B - B') \longrightarrow (\mathring{E}^n - B', B - B')$, $\tilde{g}$:$(E^n, S^{n-1}) \longrightarrow (X, X_0)$ for which the diagram

$$\begin{array}{ccc} H_q(E^n - B', B - B') & \xrightarrow{\cong} & H_q(E^n, S^{n-1} \cup B) \\ \downarrow g''_* & & \downarrow g'_* \\ H_q(\mathring{E}^n - B', B - B') & \xrightarrow{\cong} & H_q(X, X_0 \cup B) \end{array}$$

is commutative by Axiom II. Since $g'' = \text{Id}$, it follows that $g''_* = \text{Id}$, and then that g'_* is an isomorphism.

15.4. *X_0 is a deformation retract of $X_0 \cup B$.*

Proof: Define F:$(X_0 \cup B) \times I \longrightarrow X_0 \cup B$ by $F(x, t) = x$ for $x \in X$ and $F(x, t) = g((1 - t)x + tx/|x|)$ for $x \in B$. Then $F(x, 0) = x$ since $B \subset \mathring{E}^n$. When $x \subset B$, $F(x, 1) = g(x/|x|) = f(x/|x|)$. It will therefore have been proved that F is a (strong) deformation retraction when the continuity of F has been established. On $B \times I$, it is obvious that F is continuous. Since $B \times I$ is open, only the continuity of F on $X_0 \times I$ remains to be checked. Let $x_0 \in X_0$ and let $V = V_0 \cup V_1$ be an open neighborhood of x_0; that is, let V_0 be open in X_0 and $V_1 \cup f^{-1}(V_0)$ be open in E^n. It is necessary to find a neighborhood U of (x_0, t) whose image $F(U)$ lies in V. Since $V_1 \cup f^{-1}(V_0)$ is open in E^n, there is around each point $\xi \in f^{-1}(V_0)$

a small open neighborhood which contains the closed line segment from x to $x/|x|$ and whose intersection with $\mathring{E}^n$ lies entirely in V_1. Let V_1' be the union of these intersections for all $\xi \in f^{-1}(V_0)$. In the case $f^{-1}(V_0) = \varnothing$, let $V_1' = \varnothing$. Then V_1' is open in $\mathring{E}^n$, $V_1' \subset V_1$, and $V_1' \cup f^{-1}(V_0)$ is open in E^n. Let $U = (V_0 \cup V_0') \times I$; U is open. Let $(x, t) \in U$. If $x \in X_0$, then $F(x, t) = x \in V_0$. If $x \in B$; that is, if $x \in V_1'$, then $F(x, t) = g((1 - t)x + tx/|x|) \in V_0 \cup V_1'$. Hence,

$$F(U) \subset V_0 \cup V_1' \subset V_0 \cup V_1 = V,$$

and F is continuous.

15.5. Application: *The diagram*

$$\begin{array}{ccc} H_q(E^n, S^{n-1}) & \xrightarrow{i_*} & H_q(E^n, S^{n-1} \cup B) \\ \downarrow \tilde{g}_* & & \downarrow g_*' \\ H_q(X, X_0) & \xrightarrow{i_*} & H_q(X, X_0 \cup B) \end{array}$$

is commutative and the indicated homomorphisms are isomorphisms.

Proof: Let $(Y, B) \subset (X, A)$ and let $i:(Y, B) \longrightarrow (X, A)$ be the injection. Let Y be a deformation retract of X and let B be a deformation retract of A. The techniques used in 5.29 (Homotopy Axiom and Five-Lemma) are valid in the axiomatic theory and show that i_* is an isomorphism. Hence, the injections $(X, X_0) \longrightarrow (X, X_0 \cup B)$ and $(E^n, S^{n-1}) \longrightarrow (E^n, S^{n-1} \cup B)$ induce isomorphisms of the corresponding homology groups. The diagram commutes, since its homomorphisms arise from g or from injections. It was proved just before 15.4 that g_*' is an isomorphism, and the commutativity then establishes $\tilde{g}_*$ as an isomorphism.

15.6. Lemma:

$$H_q(X, X_0) \cong H_{q-1}(S^{n-1}) \text{ and } H^q(X, X_0) \cong H^{q-1}(S^{n-1}).$$

Proof: Let $\tilde{g}:(E^n, S^{n-1}) \longrightarrow (X, X_0)$ be the function induced by g. Then (in the case of augmented complexes), the first part of the lemma follows from the diagram

$$\begin{array}{ccccccc} 0 = H_q(E^n) & \longrightarrow & H_q(E^n, S^{n-1}) & \xrightarrow{\partial_*'} & H_{q-1}(S^{n-1}) & \longrightarrow & H_{q-1}(E^n) = 0 \\ & & \downarrow \tilde{g}_* & & \downarrow f_* & & \\ H_q(X) & \xrightarrow{j_*} & H_q(X, X_0) & \xrightarrow{\partial_*} & H_{q-1}(X_0) & \xrightarrow{i_*} & H_{q-1}(X) \end{array}$$

in which both ∂'_* and $\tilde{g}_*$ are isomorphisms. The proof for the cohomology groups is completed by a method analogous to that for the homology groups; it is only necessary to move the indices q and $*$ up and reverse the arrows.

Now replace the term $H_q(X, X_0)$ in the homology sequence of (X, X_0) by $H_{q-1}(S^{n-1})$, replace j_* by $\nabla = \partial'_* \tilde{g}_*^{-1} j_*$, and replace ∂_* by f_*. This yields

15.7. Lemma: *The following sequence is exact:*

$$\longrightarrow H_q(X_0) \xrightarrow{i_*} H_q(X) \xrightarrow{\nabla} H_{q-1}(S^{n-1}) \xrightarrow{f_*} H_{q-1}(X_0) \xrightarrow{i_*} .$$

Similarly, in the case of cohomology, a replacement of δ^* by f_* and the definition $\nabla = j^* \tilde{g}^{*-1}\, \delta'^*$ yields the exact sequence

$$\longleftarrow H^q(X_0) \xleftarrow{i^*} H^q(X) \xleftarrow{\nabla} H^{q-1}(S^{n-1}) \xleftarrow{f_*} H^{q-1}(X_0) \xleftarrow{i^*} .$$

When $q \neq n$ and $q \neq n - 1$, the two exact sequences of 15.7 reduce the computation of the $H_q(X)$ to that of the groups $H_q(X_0)$ because (in augmented complexes) $H_q(S^{n-1}) = 0$ and $H^q(S^{n-1}) = 0$ for $q \neq n - 1$. This proves

15.8. $H_q(X) \cong H_q(X_0)$ *and* $H^q(X) \cong H^q(X_0)$ *when* $q \neq n$ *and* $q \neq n - 1$.

When q is n or $n - 1$, the sequences become

$$0 \rightarrow H_n(X_0) \xrightarrow{i_*} H_n(X) \xrightarrow{\nabla} H_{n-1}(S^{n-1}) \xrightarrow{f_*} H_{n-1}(X_0) \xrightarrow{i_*} H_{n-1}(X) \rightarrow 0$$

$$0 \leftarrow H^n(X_0) \xleftarrow{i^*} H^n(X) \xleftarrow{\nabla} H^{n-1}(S^{n-1}) \xleftarrow{f^*} H^{n-1}(X_0) \xleftarrow{i^*} H^{n-1}(X) \leftarrow 0.$$

From the right-hand end of these sequences, it follows at once that

15.9. $$H_{n-1}(X) \cong H_{n-1}(X_0)/f_* H_{n-1}(S^{n-1}) = \operatorname{Coker} f_*,$$

and

15.10. $$H^{n-1}(X) \cong \operatorname{Ker} f^*.$$

In the case of homology, let ∇' be the function that is obtained from ∇ by restricting the image space to $\operatorname{Ker} f_*$. Then the exactness at $H_{n-1}(S^{n-1})$ implies the exactness of

15.11. $$0 \longrightarrow H_n(X_0) \xrightarrow{i_*} H_n(X) \xrightarrow{\nabla'} \operatorname{Ker} f_* \longrightarrow 0.$$

In the case of cohomology, ∇ induces an isomorphism ∇' of $H^{n-1}(S^{n-1})/f^*H^{n-1}(X_0) = \operatorname{Coker} f^*$ into $H^n(X)$. This yields the exact sequence

15.12. $$0 \longleftarrow H^n(X_0) \xleftarrow{i_*} H^n(X) \xrightarrow{\nabla'} \operatorname{Coker} f^* \longleftarrow 0.$$

Definition: A module P is called *projective* (*injective*) if and only if to each diagram which is exact at B and has the form

$$\begin{array}{ccccc} & & P & & \\ & & \downarrow & & \\ A & \longrightarrow & B & \longrightarrow & 0 \end{array} \qquad \left(\begin{array}{ccccc} & & P & & \\ & & \downarrow & & \\ A & \longleftarrow & B & \longleftarrow & 0 \end{array}\right).$$

There is a homomorphism $P \longrightarrow A$ $(P \longleftarrow A)$ for which the diagram

$$\begin{array}{ccccc} & & P & & \\ & \swarrow & \downarrow & & \\ A & \longrightarrow & B & \longrightarrow & 0 \end{array} \qquad \left(\begin{array}{ccccc} & & P & & \\ & \nearrow & \uparrow & & \\ A & \longleftarrow & B & \longleftarrow & 0 \end{array}\right)$$

is commutative.

15.13. Lemma: *Let* $0 \longrightarrow I \xrightarrow{i} A \xrightarrow{j} P \longrightarrow 0$ *be exact. If* P *is projective or* I *is injective, then* $A \cong P \oplus I$.

Proof: Suppose that P is projective. Then there is an $f: P \longrightarrow A$ such that $jf = \mathrm{Id}$. Let $a \in A$. Since $j(a - fj(a)) = 0$, there is a $b \in I$ for which $i(b) = a - fj(a)$. Application of j shows that the representation $a = i(b) + fj(a)$ is direct. The proof can be carried out similarly in the case that I is injective.

Lemma 15.13 together with Equations 15.11 and 15.12 shows that if $H_n(X_0)$ is injective or $\operatorname{Ker} f_*$ is projective (in cohomology, if $H^n(X_0)$ is projective or $\operatorname{Coker} f^*$ is injective), then

15.14.

$$H_n(X) \cong H_n(X_0) \oplus \operatorname{Ker} f_* \qquad (H^n(X) \cong H^n(X_0) \oplus \operatorname{Coker} f^*).$$

15.15. Corollary: *The isomorphisms of* 15.14 *are valid when* R *is a field. The isomorphism for homology of* 15.14 *is valid when* R *is a principal ideal ring and the coefficient module* M *is free.*

Proof:

(a) If R is a field, the modules that appear are vector spaces and hence free. Each free module is projective.

(b) If R is a principal ideal ring and the coefficient module M is free, then there is a well-known theorem according to which each submodule of M, in particular $\operatorname{Ker} f_*$, is free.

Spherical Complexes

Definition: Begin with a discrete space consisting of a finite set of points. To this attach one at a time the elements of a finite set of cells. The resulting space is called a *spherical complex*.

Since the discrete space is a compact hausdorff space, Lemmas 15.2 and 15.3 show that each spherical complex is a compact hausdorff space.

Translator's Note: Although the concept of a CW complex is not used in this book, it is worth mentioning here that the set of finite CW complexes is a proper subset of the set of spherical complexes. It is true that the finite CW complexes are also obtained by successively attaching cells to a finite set. However, their construction is subject to three further conditions: (1) For each k, all the k-cells are attached before any $(k + 1)$-cells are attached. (2) All the k-cells are attached to the $(k - 1)$-skeleton by their boundaries. (3) The interiors of the k-cells are disjoint. It is left to the reader to investigate the spherical complex which is obtained as follows: First attach a 2-cell to a point P to yield S^2 [as in Example (1) below] and then attach a 2-cell to S^2 so that the image of S^1 under the attaching function is a curve consisting of infinitely many loops in S^2 based at P. The resulting spherical complex is not a finite CW complex.

Examples:

(1) Let X_0 be a point P and $f{:}S^{n-1} \longrightarrow P$. Then f is continuous and $X = \mathring{E}_n \cup X_0$ is homeomorphic to S^n. Hence, S^n is a spherical complex.

(2) Let X_0 be an n-sphere S^n and let $f{:}S^n \longrightarrow X_0$ be the identity. Then $X = X_0 \cup E^n$ is E^n. Hence, E^n is a spherical complex.

(3) Let $X_0 = S^1$, $P \in X_0$, and let $f{:}S^0 \longrightarrow X_0$ be defined by $f(S^0) = P$. Then $X_1 = X_0 \cup E^1$ consists of two circles, a and b, which are attached at the point P.

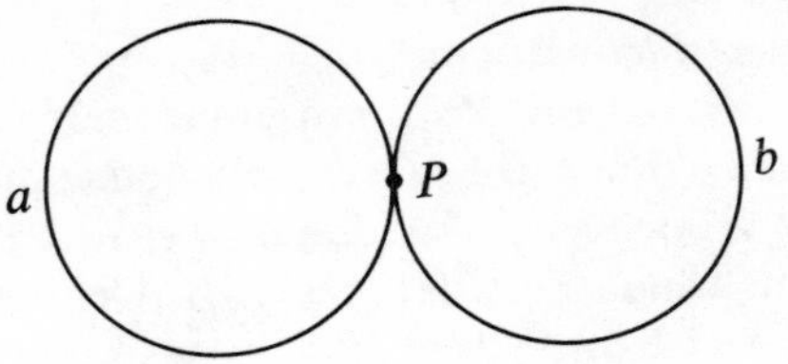

Figure 19

Now take an E^2 in the form of a rectangle. Its boundary is S^1. Map S^1 continuously into X_1 by sending the vertices onto P, one pair of opposite edges on a and the other pair of opposite edges on b. This attaches E^2 to X_1. Because of variations of orientation, there are three possiblities for this attaching:

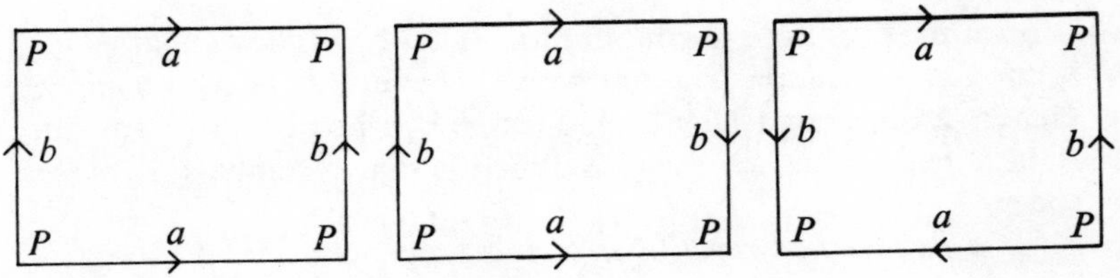

Figure 20

The first yields a torus; the second, the Kline bottle. The third is not a surface (2-manifold) since it has the local appearance of the vertex of a cone at the point P.

(4) The graphs are obtained from a finite set of points by repeated attaching of E^1. They are, therefore, spherical complexes.

(5) The projective spaces and the lens spaces will be studied later. They are spherical complexes.

(6) Any space obtained by attaching an n-cell to a spherical complex is a spherical complex.

15.16. *If X is a spherical complex, then $H_q(X) = 0$ and $H^q(X) = 0$, for $q < 0$. If m is the maximal dimension of the cells employed in the construction of X, then $H_q(X) = 0$ and $H^q(X) = 0$* for $q > m$.

Proof: The first part of the statement follows from 15.8 and the second from an induction.

15.17. The following is one way of proving that a given compact hausdorff space is a spherical complex:

Let I_1^n be the unit cube in R^n (it is a homeomorph of E^n) and let S^{n-1} be its boundary. Let $h_1:\mathring{I}_1^n \longrightarrow \mathring{I}^n$ be a homeomorphism whose image space is $\mathring{I}^n$ an open set of X. Let $h:I_1^n \longrightarrow X$ be a continuous extension of h_1 for which $\mathring{I}^n \cap h(S^{n-1}) = \varnothing$. Denote $X - \mathring{I}^n$ by X_0 and the restriction of h to S^{n-1} by $f:S^{n-1} \longrightarrow X_0$. Let the space obtained by attaching I^n to X_0 with f be denoted by X'. *The function* $\phi:X' \longrightarrow X'$ *which is defined by* $\phi(x') = x'$ *for* $x' \in X_0$ *and by* $\phi(x') = h(x')$ *for* $x' \in I_1^n$ *is a homeomorphism.*

Proof: The function ϕ is both one-to-one and onto (bijective). Let V be an open set in X, let $V \cap X_0 = V_0$, let $V \cap \mathring{I}^n = V_1$. Then $\phi^{-1}(V) \cap X_0 = V_0$ is open in X_0 and

$$f^{-1}(V_0) \cup (\phi^{-1}(V) \cap \mathring{I}_1^n)$$
$$= f^{-1}(V_0) \cup \phi^{-1}(V_0) = h^{-1}(V_0) \cup h^{-1}(V_1) = h^{-1}(V)$$

is open in X' since h is continuous. By 15.1, it follows that $\phi^{-1}(V)$ is open in X'. Hence, ϕ is continuous. The set $\mathscr{C}\mathring{I}^n = X_0$ is compact. Hence, X' is compact by 15.3. Then ϕ is a homeomorphism since it is a one-to-one, continuous function on a compact hausdorff space.

If a finite set of repetitions of this process reduces X to a finite set of points, then X is a spherical complex.

15.18. The result of 15.17 can be improved somewhat: Let $h_1:\mathring{I}_1^n \longrightarrow \mathring{I}^n$ be a homeomorphism whose image space is $\mathring{I}^n$, an open subset of X. Suppose that, for each sequence $\{x_k\} \subset \mathring{I}_1^n$ which converges to a point $x \in I_1^n$, the image sequence $\{h_1(x_k)\}$ converges to a point $\lim h_1(x_k) \in X - \mathring{I}_1^n$. Then h_1 can be extended to a continuous function $h:I_1^n \longrightarrow X$ by defining $h(x)$ to be $\lim h(x_k)$, where $x \in S^{n-1}$ and $\{x_k\}$ is a sequence in $\mathring{I}_1^n$ converging to x. The conditions of 15.17 are fulfilled, and if the resulting space X_0 is a spherical complex, then X is.

Products of Spherical Complexes

15.19. THEOREM: *If X and Y are spherical complexes, their product also is a spherical complex.* (This theorem is false for general CW complexes.)

Proof: Let X^α, $\alpha = 0, 1, \ldots, r$ and Y^β, $\beta = 0, 1, \cdots, s$ be the successive stages in the formation of X and Y from finite point sets X^0 and Y^0, respectively. Let $X^r = X$ and $Y^s = Y$. Each stage is formed from the previous by the attaching of a cell; thus $X^{\alpha+1} = X^\alpha \cup X_{\alpha+1}$ and $Y^{\beta+1} = Y^\beta \cup Y_{\beta+1}$, where the unions are disjoint and $X_{\alpha+1}$ and $Y_{\beta+1}$ are each homeomorphic to the interior of a cell of some dimension. This dimension can, of course, differ from X to Y and from one stage to another. Since X^α and Y^β are compact hausdorff spaces, so is $X^\alpha \times Y^\beta$. Let $Z_m = \cup_\alpha X^\alpha \times Y^{m-\alpha}$. Clearly, $Z_m \subset X \times Y$. As union of a finite set of compact spaces, Z_m is compact. As a compact subset of $X \times Y$, Z_m is closed in $X \times Y$. The set $X_i \times Y_{m-i}$ is a subspace of Z_m for all i. The intersections of $X_i \times Y_{m-i}$ with $X^{i-1} \times Y^{m-i}$, with $X^i \times Y^{m-i-1}$, and with $X^\alpha \times Y^{m-\alpha}$ for $\alpha \neq i$ are all empty. The complement of $X_i \times Y_{m-i}$ in Z_m is therefore

$$\begin{aligned} Z_m - X_i \times Y_{m-i} \\ &= (X^{i-1} \times Y^{m-i}) \cup (X^i \times Y^{m-i-1}) \cup \textstyle\bigcup_{\alpha \neq i} (X^\alpha \times Y^{m-\alpha}); \end{aligned}$$

hence, it is compact and closed. Therefore, $X_i \times Y_{m-i}$ is open in Z_m.

It must be proved that the space $X \times Y = Z_{r+s}$ is a spherical complex which is built up from the $X_i \times Y_j$ in a definite order. This will be proved by induction for the Z_m. The set $Z_0 = X^0 \times Y^0 = X_0 \times Y_0$ consists of finitely many points, and is therefore a spherical complex. Suppose that Z_{m-1} is known to be a spherical complex. Notice that

$$Z_m = Z_{m-1} \cup \textstyle\bigcup_{i=0}^m (X_i \times Y_{m-i}).$$

X^α is obtained from $X^{\alpha-1}$ by the attaching of a cube $I^{d(\alpha)}$ (a homeomorph of the ball $E^{d(\alpha)}$) by a continuous function $f: I^{d(\alpha)} \longrightarrow X^\alpha$ which the identity on $\mathring{I}^{d(\alpha)} = X_\alpha$ and which sends the boundary $S^{d(\alpha)-1}$ of $I^{d(\alpha)}$ into $X^{\alpha-1}$. Let $\{x_k\}$ be a sequence of points in $\mathring{I}^{d(\alpha)} = X_\alpha$ which converges to $x \in S^{d(\alpha)-1}$. Then the sequence $\{f(x_k)\} = \{x_k\}$ converges in X^α to $f(x) \in X^{\alpha-1}$ because f is continuous. In the same way, Y^β is obtained from $Y^{\beta-1}$ by attaching an $I^{e(\beta)}$ by a continuous function $g: I^{e(\beta)} \longrightarrow Y^\beta$.

Now let $\{(x_k, y_k)\}$ be a sequence of points in $X_\alpha \times Y_\beta$ that converges to a boundary point of

$$X_\alpha \times Y_\beta = \mathring{I}^{d(\alpha)} \times \mathring{I}^{e(\beta)} = \mathring{I}^{d(\alpha)+e(\beta)}.$$

Then $\{x_k\}$ converges to $x \in I^{d(\alpha)}$ and $\{y_k\}$ converges to $y \in I^{e(\beta)}$, where x or y is a boundary point. Therefore, the sequence $\{f(x_k)\} = \{x_k\}$ converges to $f(x)$ in X^α, and the sequence $\{g(y_k)\}$ converges to $g(y)$ in Y^β. Here $f(x) \in X^{\alpha-1}$ or $g(y) \in Y^{\beta-1}$ and therefore,

$$(f(x), g(y)) \in X^{\alpha-1} \times Y^\beta \cup X^\alpha \times Y^{\beta-1} \subset Z_{m-1}.$$

Now let C be Z_{m-1} or a space which is obtained from Z_{m-1} by attaching one of the sets $X_i \times Y_{m-i}$. The remaining $X_i \times Y_{m-i}$ can then be attached successively to C since the requirements of 15.18 are satisfied; namely, that the spaces involved be compact and hausdorff, that $X_i \times Y_{m-i} = \mathring{I}^{d(i)+e(m-i)}$ be open in Z_m and therefore in $C \cup (X_i \times Y_{m-i}) \subset Z_m$, and that each sequence $\{(f(x_k)g(y_k)\} = \{(x_k, y_k)\}$ converge in $C \cup (X_i \times Y_{m-i})$ to the point $(f(x), g(y)) \in Z_{m-1} \subset C$.

Thus, Z_m is obtained from Z_{m-1} by attaching the $X_i \times Y_{m-i}$ in some order. Therefore, Z_{r+s} is a spherical complex.

16

Betti Numbers and the Euler Characteristic

In the axiomatic theory, let $H_0(P) \cong R$ where R is an integral domain. Let K be the quotient field of R. Let the complexes be non-augmented. Let (X, A) be an admissible space pair. Let $\bar{H}_q(X, A)$ be defined by $\bar{H}_q(X, A) = H_q(X, A) \otimes_R K$ ($\bar{H}^q(X, A) = H^q(X, A) \otimes_R K$). Since $\otimes_R K$ preserves exactness (9.17), the same exact sequences hold for $\bar{H}^q$ and $\bar{H}_q$ as for H^q and H_q.

Definition: By the qth *betti number* $B_q(X, A)$ is meant the rank of $H_q(X, A)$; that is, the dimension of the vector space $\bar{H}_q(X, A)$ with coefficients in K. Similarly, $B^q(X, A)$ is the rank of $H^q(X, A)$.

16.1. Lemma: *Let* $\cdots \longrightarrow V_{q+1} \xrightarrow{\lambda_{q+1}} V_q \xrightarrow{\lambda_q} V_{q-1} \longrightarrow \cdots$ *be an* *exact sequence of vector spaces* V_q *of dimension* d_q, $q \in Z$. *Let all* d_{3q-1} *and* d_{3q+1} *be finite and suppose that only finitely many of them are not zero. Then all the* d_{3q} *are finite, and all but a finite set of them are zero. Furthermore,* $\sum_q^{3q}(-1)^q d_q = 0$.

Proof: Let $r_q = \dim(\text{Ker}\,\lambda_q)$. Im λ_q is isomorphic to $V_q/\text{Ker}\,\lambda_q$ and is equal to Ker λ_{q-1}. Therefore, $d_q = r_q + r_{q-1}$ for all q. This proves that

$$d_{3q} + r_{3q+1} = r_{3q} + r_{3q-1} + r_{3q+1} = d_{3q+1} + r_{3q-1}.$$

Consequently, all $d(3q)$ are finite, almost all are zero, and

$$\sum_q^{3q}(-1)^q d_q = \sum_q^{3q}(-1)^q r_q + \sum_q^{3q}(-1)^q r_{q-1} = 0.$$

Definition: Let all the ranks $B_q(X, A)$ and $B^q(X, A)$ be finite, and let at most finitely many of them be different from zero. Then $\sum_q(-1)^q B_q(X, A)$ and $\sum_q(-1)^q B^q(X, A)$ are defined and are the *euler characteristics* $\chi(X, A)$ of (X, A) in the relevant homology or cohomology theory.

16.2. THEOREM: *If two of the euler characteristics* $\chi(X, A)$, $\chi(X)$, *and* $\chi(A)$ *are defined, the third is also defined and* $\chi(X, A) + \chi(A) = \chi(X)$.

Proof: The pair (X, A) has an exact homology sequence to which Lemma 16.1 applies.

16.3. *Let* (X, X_1, X_2) *be an exact triad in which* $X = X_1 \cup X_2$ *and* $A = X_1 \cap X_2$. *If two of the numbers* $\chi(X)$, $\chi(A)$, *and* $\chi(X_1) + \chi(X_2)$ *are defined, then the third is also defined and*

$$\chi(X) + \chi(A) = \chi(X_1) + \chi(X_2).$$

Proof: Apply 16.1 to the Mayer-Vietoris sequence.

Remark: Homotopic space pairs have equal betti numbers, and therefore, equal euler characteristics. Admissible excisions preserve betti numbers and euler characteristics.

Examples:

(1) Since $R \otimes_R K = K$ and $(\sum^{\oplus} R) \otimes_R K = \sum^{\oplus} K$, it follows that $\chi(E^n) = 1$ and $\chi(S^n) = 1 + (-1)^n$.

(2) For surfaces of genus g, $B_0 = 1$, $B_1 = 2g$, $B_2 = 1$, and therefore, $\chi(F_g) = 2(1 - g)$.

16.4. THEOREM: *Let X be obtained from X_0 by attaching a cell E^n. Then $\chi(X)$ is defined if and only if $\chi(X_0)$ is defined and*

$$\chi(X) = \chi(X_0) + (-1)^n.$$

Proof: The existence of the isomorphisms

$$\tilde{g}_*: H_q(E^n, S^{n-1}) \longrightarrow H_q(X, X_0)$$
$$(\tilde{g}^*: H^q(X, X_0) \longrightarrow H^q(E^n, S^{n-1}))$$

of 15.1 shows that $\chi(X, X_0)$ is defined and that

$$\chi(X, X_0) = \chi(E^n) - \chi(S^{n-1}) = 1 - (1 + (-1)^n) = (-1)^n.$$

Then the proof can be completed by use of Theorem 16.2.

16.5. THEOREM: *Let X be a spherical complex. Let α_q be the number of q-dimensional cells employed in its construction. Then*

$$\chi(X) = \sum_q (-1)^q \alpha_q.$$

Proof: The construction begins with a hausdorff space X_0 with α_0 points. By the Direct Sum Theorem, $H_0(X_0) = \sum^{\oplus} R$ with α_0 summands. Consequently, $B_0(X_0) = \alpha_0$; and, of course, $B_q(X_0) = 0$ for $q \neq 0$. Hence, $\chi(X_0) = (-1)^0 \alpha_0$. Now the theorem follows from Theorem 16.4 by induction.

Remark: For a spherical complex X the euler characteristic $\chi(X) = \sum_q (-1)^q \alpha_q$ is independent of the homology or cohomology theory employed and always yields the same result. If X is the surface of a convex polyhedron in R^3, then $\chi(X) = \alpha_0 - \alpha_1 + \alpha_2$, where α_0 is the number of vertices, α_1 the number of edges, and α_2 the number of faces of X. Since X is homeomorphic to S^2, it follows that $\chi(X) = 2$. This yields $\alpha_0 - \alpha_1 + \alpha_2 = 2$, which is the well-known formula of Euler. If the faces are interpreted as countries on a map that is printed on a globe, Euler's formula gives some information about the possible numbers of boundary lines and intersection points of boundary lines. In geographical problems on a surface F_g of characteristic g, the corresponding equation is $\alpha_0 - \alpha_1 + \alpha_2 = 2(1 - g)$.

Examples:

(3) Let X_1 and X_2 be topological spaces containing the balls E_1^n and E_2^n, respectively. Let X be the space described in Chapter 13 obtained by adjoining X_1 to X_2 by means of the homeomorphism between E_1^n and E_2^n.

The triad $(X, X_1 - \mathring{E}_1^n, X_2 - \mathring{E}_2^n)$ is an exact triad as can be seen by means of an excision and a homotopy (since E^n lies in a larger euclidean neighborhood). By Theorem 16.3,

$$\chi(S^{n-1}) + \chi(X) = (X_1 - \mathring{E}_1^n) + (X_2 - \mathring{E}_2^n).$$

On the other hand, X_1 and X_2 are obtained from $X_1 - \mathring{E}_1^n$ and $X_2 - \mathring{E}_2^n$ by attaching cells E_1^n and E_2^n. Hence, Theorem 16.3 yields

$$1 + (-1)^{n-1} + \chi(X) = \chi(X_1) - (-1)^n + \chi(X_2) - (-1)^n$$

and then

$$\chi(X) = \chi(X_1) + \chi(X_2) - 1 - (-1)^n$$

or

$$\chi(X) = \chi(X_1) + \chi(X_2), \qquad \text{for } n > 0 \text{ and odd}$$
$$\chi(X) = \chi(X_2) - 2, \qquad \text{for } n > 0 \text{ and even.}$$

(4) The torus T and the Kline bottle K are spherical complexes in which $\alpha_0 = 0$, $\alpha_1 = 2$, and $\alpha_2 = 1$. By Theorem 16.5, $\chi(T) = \chi(K) = 0$. The adjoining of g tori to T yields F_g. Then, Example (3) shows that $\chi(F_g) = 2(1 - g)$.

(5) The real projective plane: The set of all 1-dimensional linear subspaces (lines through the origin) of R^3 is called the real projective plane P^2. To each x in S^2 associate the line through x. This is a function $S^2 \longrightarrow P^2$ which identifies antipodal points x and $-x$ of S^2. The function is made continuous by giving P^2 the quotient topology: U is open in P^2 if and only if its inverse in S^2 is open. P^2 is obtained from the northern hemisphere by identifying antipodal points of the equator. If S^1 is parametrized by means of the angle t and $f(x(t)) = x(2t)$ for $x = x(t) \in S^1$,

then $f(x) = f(-x)$, f is continuous, and f describes the identification of $x \in S^1$ with $-x \in S^1$. If E^2 is attached to S^1 by means of f, then the spherical complex so obtained is precisely P^2 with the correct topology. Hence, $\chi(P^2) = \chi(S^1) + (-1)^2 = 1$.

Definition: A topological space is called a *manifold of dimension n* if and only if each of its points has a neighborhood homeomorphic to $\mathring{E}^n$ (a euclidean neighborhood). Often there are additional hypotheses of compactness and connectedness (and therefore pathwise connectedness, since each point has a pathwise-connected neighborhood).

The function sending S^2 on P^2 in Example (5) is a local homeomorphism. Consequently, P^2 is a 2-dimensional manifold, which, as continuous image of S^2, is compact and connected.

(6) Let Φ_h be the surface which is obtained by starting with a projective plane and successively adjoining $h - 1$ more projective planes. Then

$$\begin{aligned}\chi(\Phi_h) &= \chi(\Phi_{h-1}) + 1 - 2 = \chi(\Phi_{h-r}) - r \\ &= \chi(P^2) - h - 1 = 2 - h;\end{aligned}$$

that is,

$$\chi(\Phi_h) = 2 - h.$$

16.6. THEOREM: *The euler characteristic is sufficient to distinguish all F_g and also all Φ_{2h-1} among themselves. No two of them are homotopic. On the other hand, $\chi(F_g) = \chi(\Phi_{2g})$.*

Remark: It will be proved later that the betti numbers will distinguish F_g from Φ_{2g}.

16.7. THEOREM: *If X is obtained from X_0 by adjoining an E^n by means of a continuous function $f: S^{n-1} \longrightarrow X_0$, then*

(a) $B_q(X) = B_q(X_0)$ *and* $B^q(X) = B^q(X_0)$ *when both* $q \neq n$ *and* $q \neq n - 1$.

(b) *If* $\operatorname{Ker} f_* = 0$, *then* $B_n(X) = B_n(X_0)$ *and* $B_{n-1}(X) = B_{n-1}(X_0) - 1$.

(c) *If* $\operatorname{Ker} f_* \neq 0$, *then* $B_n(X) = B_n(X_0) + 1$ *and* $B_{n-1}(X) = B_{n-1}(X_0)$.

(d) *If* $\operatorname{Coker} f^* = R$, *then* $B^n(X) = B^n(X_0) + 1$ *and* $B^{n-1}(X) = B^{n-1}(X_0)$.

(e) *If* $\operatorname{Coker} f^* \neq R$, *then* $B^n(X) = B^n(X_0)$ *and* $B^{n-1}(X) = B^{n-1}(X_0) - 1$.

Proof:

(a) Use will be made of 15.8, 15,11, and 15.12. When q is different from n and $n - 1$, the groups are isomorphic and then the betti numbers are equal. Theorem 16.4 states that $\chi(X) = \chi(X_0) + (-1)^n$. Therefore, it is sufficient to determine either the nth or the $(n - 1)$st betti number. Let $\mathscr{A} \neq 0$ be an ideal of R. Then the sequence

$$0 \longrightarrow \mathscr{A} \longrightarrow R \longrightarrow R/\mathscr{A} \longrightarrow 0$$

is exact. If $R/\mathscr{A}$ is used as torsion module, $(R/\mathscr{A}) \otimes_R K = 0$. Hence, $\mathscr{A} \otimes_R K \cong K$.

(b) From 15.11, it follows at once that $H_n(X) \cong H_n(X_0)$.

(c) Ker f_* is an ideal $\mathscr{A} \neq 0$ in R. By means of $\otimes_R K$, 15.11 shows that $B_n(X) = B_n(X_0) + 1$.

(d) An application of $\otimes_R K$ to 15.12 yields $B^n(X) = B^n(X_0) + 1$.

(e) Coker $f^* = R/\mathscr{A}$, where $\mathscr{A} = f^*(H^{n-1}(X))$ and is a non-zero ideal in R. Apply $\otimes_R K$ to 15.12. This yields $H^n(X) \cong H^n(X_0)$.

16.8. THEOREM: *If X and Y are spherical complexes, then*

$$\chi(X \times Y) = \chi(X) \cdot \chi(Y).$$

Proof: By 15.19, $X \times Y$ is a spherical complex. As in that theorem, let X and Y be built up by the successive attaching of the cells X_i and Y_j respectively. Let the number of the X_i and Y_j of dimension q be $\alpha(q)$ and $\beta(q)$, respectively. Then in 15.19 it was proved that $X \times Y$ was built up by attaching of all the cells $X_i \times Y_j$. Let $\gamma(q)$ be the number of q-dimensional cells among these. Then $\gamma(q) = \sum_{i+j=q} \alpha(i)\beta(j)$ and consequently,

$$(X \times Y) = \sum_q (-1)^q \gamma(q) = \sum_{i,j} (-1)^{i+j} \alpha(i)\beta(j) = \chi(X) \cdot \chi(Y).$$

Examples:

(1) The n-dimensional torus T^n is the cartesian product of n copies of S^1. For T^n, it follows that $\chi(T^n) = (\chi(S^1))^n = 0$.

(2) $$\chi(S^r \times S^s) = (1 + (-1)^r)(1 + (-1)^s) = \begin{cases} 4 & \text{if both } r \text{ and } s \text{ are even} \\ 0 & \text{otherwise.} \end{cases}$$

Among other things, it follows from $\chi(S^2 \times S^2) = 4$ that $S^2 \times S^2$ is not homotopic to any cell or sphere.

(3) Since $\chi(P^2) = 1$, $\chi(P^2 \times P^2) = 1$.

17

Complex and Real Projective Spaces

Let V be an $(n+1)$-dimensional vector space over the field C (R) of real (complex) numbers with the product topology. The vectors will be written $z = (z_0, z_1, \ldots, z_n)$, and $|z|$ will be defined by $|z| = |z_0|^2 + \cdots + |z_n|^2$. This defines a metric $|z - z^1|$ on V whose topology is consistent with the product topology. The set of all 1-dimensional linear subspaces of V (lines through the origin) is called the *n-dimensional complex* (*real*) *projective space* and is denoted by CP^n (P^n). If V^* is the set $\{z \mid z \in V, z \neq 0\}$, let the function $\phi: V^* \longrightarrow CP^n$ $(\phi: V \longrightarrow P^n)$ be defined by

$$\phi(z) = \text{line through } z = \{\lambda z \mid \lambda \in C (\lambda \in R)\}.$$

The projective space is given the quotient topology induced by ϕ, that is, the largest topology in which ϕ is continuous. Here U is open in CP^n (P^n) if and only if $\phi^{-1}(U)$ is open in V^*.

Let $S^{2n+1} = \{z \mid z \in V, |z| = 1\}$ $(S^n = \{z \mid z \in V, |z| = 1\})$. The restriction $\tilde{\phi}: S^{2n+1} \longrightarrow CP^n$ $(\tilde{\phi}: S^n \longrightarrow P^n)$ of ϕ to S^{2n+1} (to S^n) is continuous because S^{2n+1} (S^n) is a subspace of V^* and onto because $\phi(z) = \phi(z/|z|)$. When z and z' are in S^{2n+1} (S^n), then $|z| = |z'| = 1$; and $\tilde{\phi}(z) = \tilde{\phi}(z')$ if and only if $z = e^{i\alpha}z'$ for some real $\alpha(z = \pm z')$. Let P and P_1 be two points of CP^n (P^n). Their inverse images $\tilde{\phi}^{-1}(P)$ and $\tilde{\phi}^{-1}(P_1)$ are circles on S^{2n+1} (pairs of points on S^n). Consequently, the two inverse images are disjoint closed sets in S^{2n+1} (S^n). Since S^{2n+1} (S^n) is a normal space, the inverses can be separated by open sets V_1 and V_2. The cones from the origin on V_1 and V_2 are disjoint open sets in V^*. Hence, $\phi(V_1)$ and $\phi(V_2)$ are open and disjoint and CP^n (P^n) is a hausdorff space. As the continuous image of a compact set, CP^n (P^n) is compact.

17.1. THEOREM: *CP^n (P^n) is a spherical complex obtained from CP^{n-1} (P^{n-1}) by attaching the ball E^{2n} (E^n).*

Proof: Embed the n-dimensional vector space V_n in V by associating the vector $u = (u_1, u_2, \cdots, u_n)$ with the vector $(0, u) = (0, u_1, u_2, \cdots, u_n) \in V$. The vectors $u \in V_n$ for which $|u| \leq 1$ form an E^{2n} (E^n) with boundary S^{2n-1} (S^{n-1}). Define the function $g: E^{2n} \longrightarrow S^{2n+1}$ $(g: E^n \longrightarrow S^n)$ by $g(u) = (\sqrt{1 - |u|}, u)$. Obviously $|g(u)| = 1$, g is continuous, and $g(u)$ has a real, non-negative first coordinate. Let $f = \tilde{\phi}g$, $f: E^{2n} \longrightarrow CP^n$ $(f: E^n \longrightarrow P^n)$. Then f is continuous. Let $P \in CP^n$ $(P \in P^n)$. Among all z for which $\tilde{\phi}(z) = P$ there is precisely one, say $(z_0, z_1, \cdots, z_n)$ for which z_0 is real and non-negative. Therefore, $z_0 = \sqrt{1 - |z|}$ and $P = \tilde{\phi}g(z_1, \cdots, z_n)$. Hence, f is surjective. If $u \in \mathring{E}^{2n}$ and $v \in E^{2n}$ $(u \in \mathring{E}^n$ and $v \in E^n)$, then $|u| < 1$ and $|v| \leq 1$. From $f(u) = f(v)$, it follows that $(\sqrt{1 - |u|}, u) = (\sqrt{1 - |v|}, v)$ and therefore, $(\sqrt{1 - |u|}, u) = e^{i\alpha}(\sqrt{1 - |v|}, v)$ which is possible only for $e^{i\alpha} = 1$ and $u = v$. Hence, $f(S^{2n-1}) \cap f(E^{2n}) = \varnothing$ $(f(S^{n-1}) \cap f\ (E^n) = \varnothing)$ and the restriction of f to $\mathring{E}^{2n}$ (E^n) is one-to-one. Let W be an open set in $\mathring{E}^{2n}$; $E^{2n} - W$ is compact, and therefore has a compact image $f(E^{2n} - W)$. Since $f(W) = f(E^{2n}) - f(E^{2n} - W)$, $f(W)$ is open in $f(E^{2n})$ and thus f is an open mapping on $\mathring{E}^{2n}$. Hence, the restriction of f to $\mathring{E}^{2n}$ is a homeomorphism.

It was just proved that $f(S^{2n-1}) \cap f(\mathring{E}^{2n}) = \varnothing$ $(f(S^{n-1}) \cap f(\mathring{E}^n) = \varnothing)$. When $|u| = 1$, then $g(u) = (0, u)$ and $f(u) = \tilde{\phi}(0, u)$. This shows that $f(S^{2n-1})$ is the CP^{n-1} formed in the V_n that was embedded in V ($f(S^n)$ is the P^{n-1} formed in the V_n that was embedded in V). As a compact subset of a hausdorff space, $f(S^{2n-1}) = CP^{n-1}$ is closed in CP^n ($f(S^{n-1}) = P^{n-1}$ is closed in P^n). Hence, $\mathscr{C}CP^{n-1} = f(\mathring{E}^{2n})$ is open in CP^n ($\mathscr{C}P^{n-1} = f(E^n)$ is open in P^n). Consequently, the conditions of 15.17 are satisfied, and CP^n is obtained from CP^{n-1} (P^n is obtained from P^{n-1}) by attaching E^{2n} (E^n).

The points of the set $f(\mathring{E}^{2n})$ ($f(\mathring{E}^n)$) have a euclidean neighborhood in this set of dimension $2n$ (n). Since each point of a projective space can be sent into any other by a projective transformation, and thus by a homeomorphism, CP^n is a $2n$-dimensional manifold (P^n is an n-dimensional manifold) which, as continuous image of S^{2n+1} (S^n), is compact and connected.

Exercises:

(a) CP^0 is a point; P^0 is a point.

(b) P^1 is a circle S^1.

(c) CP^1 is a two-sphere S^2.

Groups of the Complex Projective Space CP^n

If $n = 0$, CP^n is a single point. For the computation of $H_q(CP^n)$ and $H^q(CP^n)$ use is made of 15.8 to 15.12. Since CP^n is built up of cells of dimension $\leq 2n$, it follows that $H_q(CP^n) = 0$ and $H^q(CP^n) = 0$ for $q > 2n$ and for $q < 0$. This proves that Coker $f_* = 0$ and Ker $f^* = 0$ since $H_{2n-1}(CP^{n-1}) \cong H^{2n-1}(CP^{n-1}) = 0$. Formulas 15.9 and 15.10 then yield $H_{2n-1}(CP^n) = 0$ and $H^{2n-1}(CP^n) = 0$. Furthermore, Ker $f_* = H_{2n-1}(S^{2n-1})$ and Coker $f^* = H^{2n-1}(S^{2n-1})$. Hence, 15.11 yields

$$H_{2n}(CP^n) \cong H_{2n-1}(S^{2n-1}) \cong M,$$

and 15.12 yields

$$H^{2n-1}(CP^n) \cong H^{2n-1}(S^{2n-1}) \cong M.$$

When q is different from $2n$ and $2n - 1$, 15.8 shows that $H_q(CP^n) \cong H_q(CP^{n-1})$ and $H^q(CP^n) \cong H^q(CP^{n-1})$. This proves for the cases $0 < 2q < 2n$ that

$$H_{2q}(CP^n) \cong H_{2q}(CP^{n-1}) \cong \cdots \cong H_{2q}(CP^q) \cong M$$

and

$$H_{2q-1}(CP^n) \cong H_{2q-1}(CP^q) = 0.$$

In the same manner, it follows that $H^{2q}(CP^n) \cong M$ and $H^{2q-1}(CP^n) = 0$.

These results can be summarized as

17.2. THEOREM: *For the complex projective space CP^n*

$$H_q(CP^n) \cong H^q(CP^n) = 0 \text{ for } q < 0, q > 2n, q \text{ odd},$$
$$q = 0 \text{ (augmented)},$$
$$H_q(CP^n) \cong H^q(CP^n) \cong M \text{ for } q \text{ even with } 0 < q \leq 2n, \text{ and for}$$
$$q = 0 \text{ (non-augmented)}.$$

17.3. Corollary: *If $M = R$ and M is an integral domain, then* $\chi(CP^n) = n + 1$.

Groups of the Real Projective Space P^n

By Theorem 17.1, P^n is a spherical complex formed by attaching E^n to P^{n-1}. The attaching function was precisely the canonical mapping of S^{n-1} on P^{n-1}.

For the computation of $H_n(P^n)$ and $H_{n-1}(P^n)$, let α be the antipodal function $\alpha(x) = -x$ on S^n. The set E^n_+ will be designated by E and the set E^n_- by αE. Consider the diagram (top of p. 179). The indicated homomorphisms are boundary operators, or else are induced by the canonical function $f: S^n \longrightarrow P^n$ or by injections. Consequently, all triangles and the two rectangles are commutative. The antipodal function α induces the homomorphisms

$$\alpha_0: (S^n, S^{n-1}) \longrightarrow (S^n, S^{n-1}); \alpha_1: (\alpha E, S^{n-1}) \longrightarrow (E, S^{n-1}); \alpha_2: (S^n, \alpha E) \longrightarrow (S^n, E).$$

The injections on the right side of the diagram have been written in terms of the injections on the left side and α. The symbol * has been omitted from all homomorphisms in the diagram and the following computation. By definition, the triad $(S^n, E, \alpha E)$ is exact. Therefore, k_1 and $\alpha_2 k_1 \alpha_1$ are isomorphisms. The two diagonals are exact because they are parts of homology sequences. By the Hexagon Theorem (13.1), it follows that $f_1 j = f_2 k^{-1} l + f_2 k^{-1} l \alpha$. By the example to Theorem 7.16 (which also holds in the axiomatic theory), $\alpha = (-1)^{n+1}$. Therefore,

17.4. $$h_2 f = f_1 j = (1 + (-1)^{n+1}) f_2 k^{-1} l.$$

Here, the function f_2 is the isomorphism $\tilde{g}_*$ of 15.5 (except for an

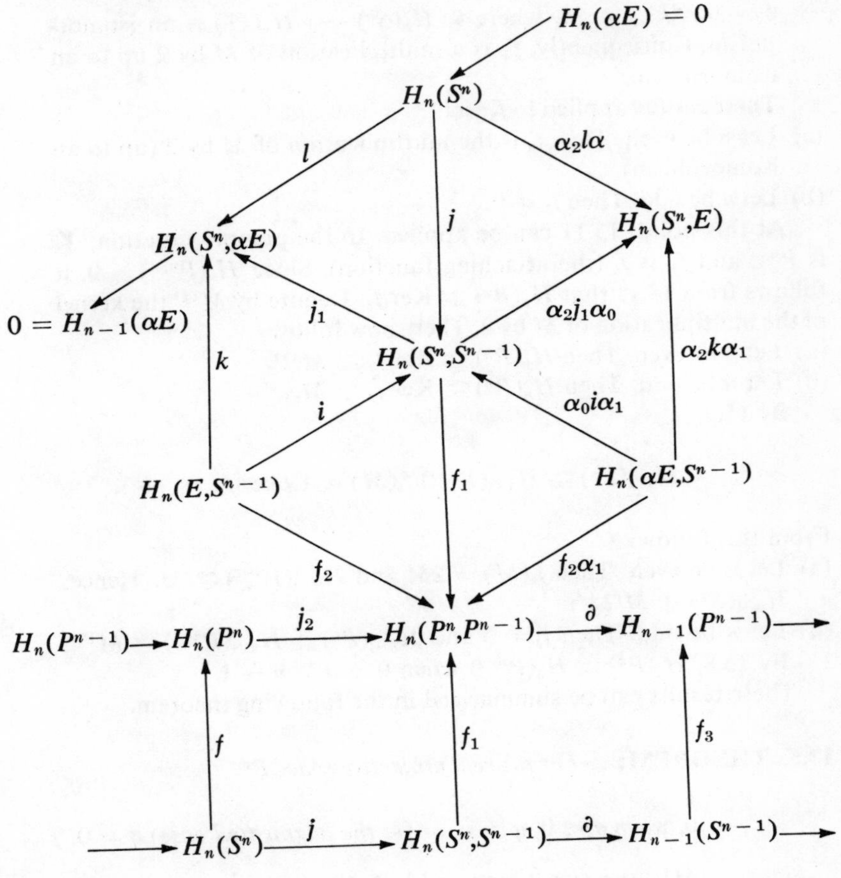

isomorphism induced by the homomorphism $E \longrightarrow E^n$). Consequently, f_2 is an isomorphism. The homology sequence of $(S^n, \alpha E)$ shows, since $H_q(\alpha E) = 0$, that l also is an isomorphism. Since i is the injection, $f_2 = f_1 i$. Consequently, f_1 is an epimorphism. Now consider 17.4.

(a) Let n be even. Since P^{n-1} is built up of cells of dimension at most $n - 1$, $H_n(P^{n-1}) = 0$ and j_2 is a monomorphism. From 17.4, it then follows that $f = 0$.

(b) Let n be odd. Then $n - 1$ is even and $f_3 = 0$ by (a). Therefore, $\partial f_1 = 0$. Since f_1 is an epimorphism, it follows that $\partial = 0$. Hence, j_2 is an isomorphism. From 17.4, it now follows that

$f = 2j_2^{-1}f_2k^{-1}l = 2\theta$, where $\theta: H_n(S^n) \longrightarrow H_n(P^n)$ is an isomorphism. Consequently, f_* is a multiplication of M by 2 up to an isomorphism.

These results applied to f_3 yield:

(a) Let n be even. Then f_3 is the mutliplication of M by 2 (up to an isomorphism).

(b) Let n be odd. Then $f_3 = 0$.

At this point, 15.11 can be applied. In the present situation, X_0 is P^{n-1} and f_* is f_3 (the attaching function). Since $H_n(P^{n-1}) = 0$, it follows from 15.11 that $H_n(P^n) \cong \operatorname{Ker} f_3$. Denote by $M^{(2)}$ the kernel of the multiplication of M by 2. There now follow

(a) Let n be even. Then $H_n(P^n) \cong \operatorname{Ker} f_3 \cong M^{(2)}$.

(b) Let n be odd. Then $H_n(P^n) \cong \operatorname{Ker} f_3 \cong M$.

By 15.9,

$$H_{n-1}(P^n) \cong H_{n-1}(P^{n-1})/f_3(M) = \operatorname{Coker} f_3.$$

From this follows

(a) Let n be even. Then $f_3(M) = 2M$ and $H_{n-1}(P^{n-1}) \cong M$. Hence, $H_{n-1}(P^n) \cong M/2M$.

(b) Let n be odd. Then $f_3 = 0$ and $H_{n-1}(P^n) \cong H_{n-1}(P^{n-1}) \cong M^{(2)}$. By 15.8, $H_q(P^n) \cong H_q(P^{n-1})$ when $0 < q < n - 1$.

These results can be summarized in the following theorem.

17.5. THEOREM: *For the real projective space P^n*

$$H_q(P^n) \cong \begin{cases} 0, \text{ when } q < 0, q > n, \text{ or (in the augmented case) } q = 0. \\ M^{(2)}, \text{ when } q \text{ is even and } 1 \leq q \leq n. \\ M/2M, \text{ when } q \text{ is odd and } 1 \leq q \leq n - 1. \\ M, \text{ when } q = n \text{ is odd or when (non-augmented) } q = 0. \end{cases}$$

17.6. Corollary: $\chi(P^n) = \sum(-1)^q = \begin{cases} 0, \text{ when } n \text{ is odd} \\ 1, \text{ when } n \text{ is even} \end{cases}$ *provided that $M = R$ and M is an integer domain.*

The cohomology groups $H^q(P^n)$ are obtained analogously. In their computation, the direction of the arrows and the order of the products in the diagram need be reversed. This yields the following theorem.

17.7. THEOREM: *The groups $H^q(P^n)$ are obtained from the list for $H_q(P^n)$ by interchanging the rôle of $M^{(2)}$ and $M/2M$.*

The case in which $M = Z$ (the integers) is of interest. In this case, $M^{(2)} = 0$ and $M/2M$ is a group of order 2. Thus, the projective spaces P^n, $n > 1$, furnish examples of spaces in which some homology groups have torsion; that is, they have an element $a \neq 0$ for which there is an integer r such that $ra = 0$.

18

Maps of S^n on S^n and Lens Spaces

18.1. Lemma: *Let (X, X_1, X_2) be an exact triad in which $X = X_1 \cup X_2$ and $A = X_1 \cap X_2$. Suppose that three functions f, f_1, f_2: $(X, A) \longrightarrow (Y, B)$ have the following properties:*

18.2.
$$\begin{cases} f_1(x) = f(x), \text{ when } x \in X_1, \\ f_2(x) = f(x), \text{ when } x \in X_2, \\ f_1(X_2) \subset B, \\ f_2(X_1) \subset B. \end{cases}$$

Then $f_* = f_{1*} + f_{2*}$ and $f^* = f_1^* + f_2^*$.

Proof: Consider the following diagram (in which the symbol * has been deleted throughout):

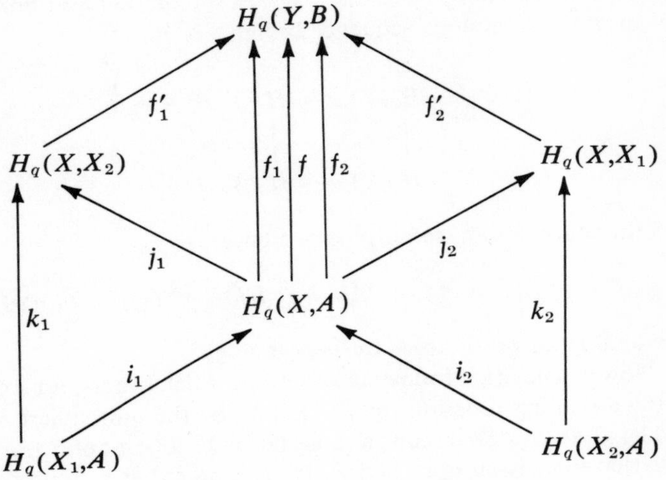

By 18.2, the functions f_ν' that are induced by f_ν are well defined and $f_\nu' j_\nu = f_\nu$, $\nu = 1, 2$. The subdiagram

$$\begin{array}{ccc} H_q(X, X_2) & & H_q(X, X_1) \\ \uparrow \quad \nwarrow \; H_q(X, A) \; \nearrow \quad \uparrow \\ H_q(X_1, A) \; \nearrow & & \nwarrow \; H_q(X_2, A) \end{array}$$

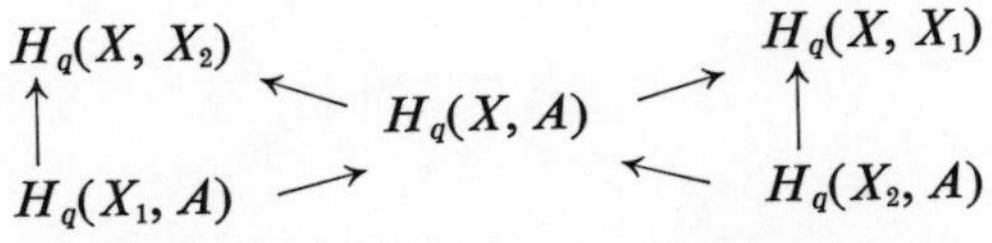

is a part of the Mayer-Vietoris diagram. Since the triad is exact, k_1 and k_2 are isomorphisms. The diagonals are exact, and the two triangles are commutative. Therefore, Lemma 10.4 applies, and it follows that $\mathrm{Id} = i_1 k_1^{-1} j_1 + i_2 k_2^{-1} j_2$; hence, $f = f i_1 k_1^{-1} j_1 + f i_2 k_2^{-1} j_2$. Since f and f_ν coincide on X_ν, it follows that

$$f i_\nu = f_\nu i_\nu = f_\nu' j_\nu i_\nu = f_\nu' k_\nu.$$

Hence,

$$f = f_1' j_1 + f_2' j_2 = f_1 + f_2.$$

The analogue in cohomology is

$$f = j_1 k_1^{-1} i_1 f + j_2 k_2^{-1} i_2 f = j_1 f_1' + j_2 f_2' = f_1 + f_2.$$

18.2a. Corollary: *If $H_q(B) = 0$ $(H^q(B) = 0)$ for all q, and if $\bar{f}, \bar{f}_1, \bar{f}_2 : X \longrightarrow Y$ are the three functions induced by $f, f_1,$ and f_2, respectively, then $\bar{f}_* = \bar{f}_{1*} + \bar{f}_{2*}$ and $\bar{f}^* = \bar{f}_1^* + \bar{f}_2^*$.*

Proof: Since $H_q(B) = 0$, the diagram for the induced homomorphism of the homology sequence becomes

$$\begin{array}{ccccccc} 0 & \longrightarrow & H_q(Y) & \xrightarrow{j} & H_q(Y, B) & \longrightarrow & 0 \\ & & \uparrow \bar{f}, \bar{f}_1, \bar{f}_2 & & \uparrow f, f_1, f_2 & & \\ \cdots & \longrightarrow & H_q(X) & \xrightarrow{j_3} & H_q(X, A) & \longrightarrow & \cdots \end{array}$$

and therefore, j is an isomorphism. Hence,

$$\bar{f} = j^{-1}fj_3 = j^{-1}(f_1 + f_2)j_3 = j^{-1}f_1j_3 + j^{-1}f_2j_3 = \bar{f}_1 + \bar{f}_2.$$

The analogous proof holds for cohomology.

Now consider the following situation, which occurs, for example, in the adjoining function for P^2: Let S^n be the unit sphere in R^{n+1} and let $f: S^n \longrightarrow S^n$ be continuous. Let x_0 be a point of S^n, and suppose that there is an open ball $\mathring{E}_1 \subset S^n$ centered at x_0 for which the inverse image $f^{-1}(\mathring{E}_1) = \bigcup_{\nu=1}^{r} F'_\nu$ is the disjoint union of r open sets $F'_r \subset S^n$ each of which is homeomorphic to $\mathring{E}_1$ by a restriction of f.

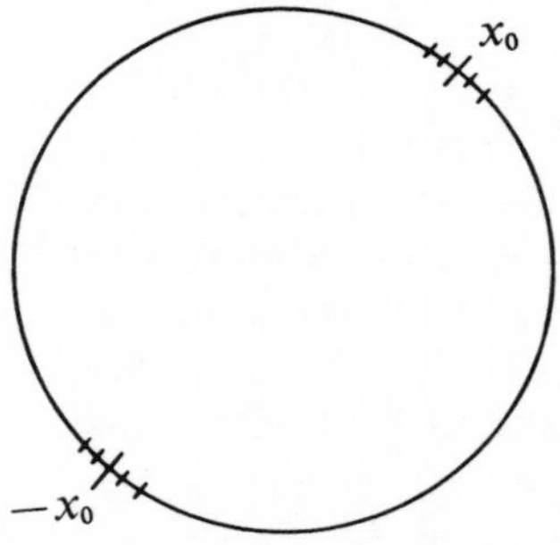

Figure 21

Choose a closed ball $E \subset \mathring{E}_1$ centered at x_0. Let $F_\nu = f^{-1}(E) \cap F'_\nu$. Since f yields a homeomorphism of F'_ν on $\mathring{E}_1$, F_ν is closed.

18.3. Lemma: *The triads $(S^n, F_\nu, \mathscr{C}\mathring{F}_\nu)$ are exact.*

Proof: Choose an open ball U around x_0 such that $\overline{U} \subset \mathring{F}_\nu$. Then $(S^n - U, F_\nu - U) \longrightarrow (S^n, F_\nu)$ is an excision by the Excision Axiom. Furthermore, $(S^n - U, F_\nu - U)$ is homotopic to $(\mathscr{C}\mathring{F}_\nu, F_\nu \cap \mathscr{C}\mathring{F}_\nu)$.

Hence, $H_q(S^n, F_\nu) \cong H_q(\mathscr{C}\mathring{F}_\nu, F_\nu \cap \mathscr{C}\mathring{F}_\nu)$. The other isomorphism is obtained in the same way.

Let y_0 be the point of S^n antipodal to x_0. To each λ for which $0 \leq \lambda \leq 1$, define a function $h_\lambda: S^n \longrightarrow S^n$ as follows: Let μ be half of a great circle passing through x and leading from x_0 to y_0 (this is unique when $x \neq x_0$ and $x \neq y_0$). The image $h_\lambda(x)$ will lie on μ. Let P be the point of intersection of μ with the boundary of E. If x lies on the arc from P to y_0, the distance of $h_\lambda(x)$ from y_0 shall be λ times the distance from x to y_0. If x lies on the arc from x_0 to P, then $h_\lambda(x)$ shall divide the arc from x_0 to $h_\lambda(P)$ in the same ratio as x divides the arc from x_0 to P. The function $h_\lambda(x)$ satisfying these conditions is well defined, and the function $H: S^n \times I \longrightarrow S^n$ defined by $H(x, \lambda) = h_\lambda(x)$ is continuous. Obviously, $h_1 = \text{Id}$, h_0 is a homeomorphism of $\mathring{E}$ on $\mathscr{C}(y_0)$, $h_0(\mathscr{C}\mathring{E}) = y_0$, and Id is homotopic to h_0. Define g by $g = h_0 f$. Then f and g are homotopic. If $x \notin \bigcup_{\nu=1}^r \mathring{F}_\nu$, then $f(x) \notin \mathring{E}$ and $g(x) = y_0$. Define $g_\nu: S^n \longrightarrow S^n$ by

$$g_\nu(x) = \begin{cases} g(x) \text{ if } x \in F_\nu \\ y_0 \text{ if } x \in \mathscr{C}\mathring{F}_\nu. \end{cases}$$

Remember that F_ν and $\mathscr{C}\mathring{F}_\nu$ are closed and that if $x \in F_\nu \cap \mathscr{C}\mathring{F}_\nu$, then $x \notin \bigcup_\nu \mathring{F}_\nu$ and $g(x) = y_0$. Hence, by Lemma 5.16, g_ν is continuous. Define $\psi: S^n \longrightarrow S^n$ by

$$\psi(x) = \begin{cases} g(x) \text{ if } x \in \bigcup_{\nu=2}^{r} F_\nu \\ y_0 \text{ if } x \in \mathscr{C} \bigcup_{\nu=2} F_\nu. \end{cases}$$

If $x \in \bigcup_{\nu=2}^r F_\nu \cap \mathscr{C} \bigcup_{\nu=2}^r \mathring{F}_\nu$, then $x \notin \bigcup_{\nu=1}^r \mathring{F}_\nu$ since the F_ν are disjoint, and $g(x) = y_0$. Then ψ is continuous by Lemma 5.16.

The functions g, g_1, and ψ induce functions $(S^n, F_1 \cap \mathscr{C}\mathring{F}_1) \longrightarrow (S^n, y_0)$ to which Lemma 18.1 will now be applied: By definition, $g_1(x) = g(x)$ if $x \in F_1$ and $g(x) = y_0$ for $x \in \mathscr{C}\mathring{F}_1$. When $x \in F_1$, $\psi(x) = y_0$. When $x \in \mathscr{C}\mathring{F}_1$, either $x \notin \bigcup_{\nu=2}^r \mathring{F}_\nu$ and then $x \notin \bigcup_{\nu=2}^r F_\nu$ which in turn implies that $\psi(x) = y_0 = g(x)$, or else $x \in \bigcup_{\nu=2}^r F_\nu$ and then $\psi(x) = g(x)$. In either case, $\psi(x) = g(x)$ when $x \in \mathscr{C}\mathring{F}_1$. This means that the hypotheses of Lemma 18.1 are satisfied. Even Corollary 18.2 holds, since $H_q(y_0) = H^q(y_0) = 0$ (in augmented complexes). Hence, $g_* = g_{1*} + \psi_*$ and $g^* = g_1^* + \psi^*$.

An induction on r yields $g_* = \sum_{\nu=1}^{r} g_{\nu*}$ and $g^* = \sum_{\nu=1}^{r} g_\nu^*$. Since g and f are homotopic, it follows that

18.4. $$f_* = \sum_{\nu=1}^{r} g_{\nu*} \text{ and } f^* = \sum_{\nu=1}^{r} g_\nu^*.$$

Actually, each g_ν is homotopic to an orthogonal transformation of the sphere, and therefore, $g_{\nu*} = \pm 1$. However, a proof of this is beyond the scope of this book. Two simple conditions that imply that $g_{\nu*} = \pm 1$ are as follows:

18.5.

(1) There are continuous functions $f_\nu: S^n \longrightarrow S^n$ which are homotopic to an orthogonal transformation of S^n and for which both $f_\nu(x) = f(x)$ when $x \in \mathring{F}_\nu$, and $f_\nu(x) \in \mathscr{C}E$ when $x \notin \mathring{F}_\nu$. Then $g = h_0 f$, and both $g_{\nu*} = f_{\nu*} = \pm 1$ and $g_\nu^* = \pm 1$ hold for every coefficient module M in the axiomatic theory.

(2) There are continuous functions $f_\nu: S^n \longrightarrow S^n$ which are homotopic to a homeomorphism of S^n onto S^n and for which both $f_\nu(x) = f(x)$ when $x \in \mathring{F}_\nu$, and $f_\nu(x) \in \mathscr{C}\mathring{E}$ when $x \notin \mathring{F}_\nu$. Again, it follows that $g_{\nu*} = f_{\nu*}$. It has been proved that $f_{\nu*}$ represents an automorphism of $H_n(S^n) \cong M$. In the case $M = Z$, it follows that $f_{\nu*} = \pm 1$ and $f_\nu^* = \pm 1$.

Let S^1 be interpreted as R/Z (S^1 is parametrized by a central angle) and let $n \in Z$. Multiplication of the elements of R by n induces a homomorphism $\theta_n: S^1 \longrightarrow S^1$ which is defined by $\theta_n(x) = nx$.

18.6. *In the axiomatic theory, the homomorphism θ_{n*} is a multiplication of M by n for any coefficient module M. Similarly, $\theta_n^* = n$.*

Proof: The conclusion follows readily when $n = 0$. Since θ_{-1} is an orthogonal transformation, it follows that $\theta_{-1*} = -1$. The relation $\theta_{-n} = \theta_{-1}\theta_n$ permits consideration to be centered on positive n. Let I be the interval $0 < x < \epsilon$, where ϵ is sufficiently small. Then $\theta_n^{-1}(I)$ consists of the disjoint sets $I_\nu = (1/n)I + (\nu/n)$, $\nu = 0, 1, \cdots, n - 1$. From 18.5 (1), it is sufficient to extend the restriction θ_n/I_ν to a continuous function $h: S^1 \longrightarrow S^1$ that is homotopic to the identity. Application of a rotation (which obviously is homotopic to the identity) shows that only I_0 need be considered. Figure 22 yields such an h and the required homotopy.

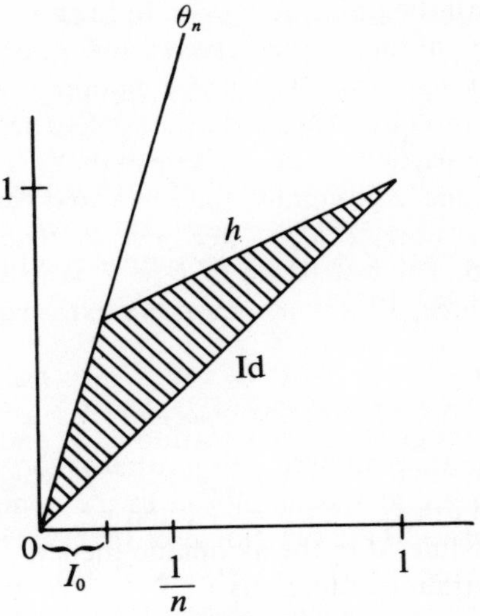

Figure 22

Application to Lens Spaces

Let $X_1 = S^1$ be a circle. Attach E^2 to X_1 by means of the function $\theta_n\colon S^1 \longrightarrow X_1$ which multiplies angles by $n \in Z_1$. Denote by $M^{(n)}$ the kernel of the multiplication by n of M. It was just proved that $\operatorname{Ker} \theta_{n*} = M^{(n)}$. Let $X_2 = X_1 \cup \mathring{E}^2$ be the space obtained by the attaching. The exactness of sequence 15.11 shows that

$$0 \longrightarrow H_2(X_1) \longrightarrow H_2(X_2) \longrightarrow \operatorname{Ker} \theta_n \longrightarrow 0$$

is exact. From $H_2(X_1) = 0$, it follows that $H_2(X_2) \cong M^{(n)}$. From 15.9, it follows that $H_1(X) \cong M/(nM)$. From 15.10 and 15.12, it follows that $H^2(X_2) \cong M/(nM)$ and $H^1(X_2) \cong M^{(n)}$.

The space X_2 is homeomorphic to a space X_2^1 which will now be constructed in a somewhat different manner: Let m and n be two relatively prime integers. Let S^2 be given coordinates (λ, ψ), where λ is the product of the geographic longitude by $1/2\pi$ and ψ is the latitude. The points of S^2 are now divided into equivalence classes such that when $\psi > 0$, the equivalence class of (λ, ψ) contains only the additional point $(\lambda + m/n, -\psi)$, and when $\psi = 0$, the class of $(\lambda, 0)$ consists of all the points $(\lambda + \nu/n, 0)$ for $\nu = 0, 1, \cdots, n-1$. Let X_2^1 be the set of equivalence classes and $F: S^2 \longrightarrow X_2^1$ be the function sending points into the class containing them. Let X_2^1 be given the identification topology induced by F. Then F is open and continuous and, in particular, $F(\mathring{E}_+^2)$ is open in X_2^1. The restriction of F to $\mathring{E}_+^2$ is one-to-one, continuous, and open, and hence, is a homeomorphism. Furthermore, $F(S^1) \cap F(\mathring{E}_+^2) = \varnothing$. Thus, condition (a) of 15.17 is fulfilled. The restriction of F to S^1 is just the multiplication of S^1 by n. Hence, X_2^1 is homeomorphic to the space X_2.

18.7. THEOREM: *For $F: S^2 \longrightarrow X_2$, the induced homomorphisms F_* and F^* are both zero homomorphisms.*

Proof: The diagram employed here is almost the same as the one which precedes 17.4 and was employed in the computation of the groups of P^n. In place of the function α of that diagram, the present proof makes use of the function $\alpha: S^2 \longrightarrow S^2$ defined by $\alpha((\lambda, \psi)) = (\lambda + m/n, -\psi)$. This is an orthogonal transformation whose determinant is -1. Consequently, $\alpha_* = -1$ $(\alpha^* = -1)$. Let E be E_+^2, then $\alpha E = E_-^2$. The functions induced by α and F are distinguished by their subscripts; in particular, $\alpha_0: (S^2, S^1) \longrightarrow (S^2, S^1)$, $\alpha_1: (E, S^1) \longrightarrow (\alpha E, S^1)$, and $\alpha_2: (S^2, \alpha E) \longrightarrow (S^2, E)$. The functions i, j, j_1, k, l are injections. As before, the injections on the right side are obtained by means of α from those on the left.

The lower left-hand triangle is trivially commutative. On the lower right-hand side, commutativity is a consequence of $F_{1*}\alpha_{0*}i_* = F_{2*}$, which is proved as follows: Let $x = (\lambda, \psi) \in E$; i.e. $\psi \geq 0$. Then

$$F_1\alpha_0 i(\lambda, \psi) = F_1(\lambda + m/n, -\psi)$$
$$= \left\{ \begin{array}{l} (\lambda, \psi) \cup (\lambda + m/n, -\psi), \text{ if } \psi > 0 \\ \bigcup_{\nu=0}^{n-1} (\lambda + \nu/n, 0), \text{ if } \psi = 0 \end{array} \right\} = F_2(x).$$

The triad $(S^n, E, \alpha E)$ is exact. Hence, k_* and also $\alpha_{2*}k_*\alpha_*^{-1}$ are isomorphisms. All of the hypotheses of the Hexagon Theorem (13.1) are satisfied. This proves that

$$F_{1*}j_* = F_{2*}k_*^{-1}l_* + F_{2*}k_*^{-1}l_*\alpha_*^{-1}.$$

Since $\alpha_* = -1$, it follows that $F_{1*}j_* = 0$, and hence, $j_{2*}F_* = 0$.

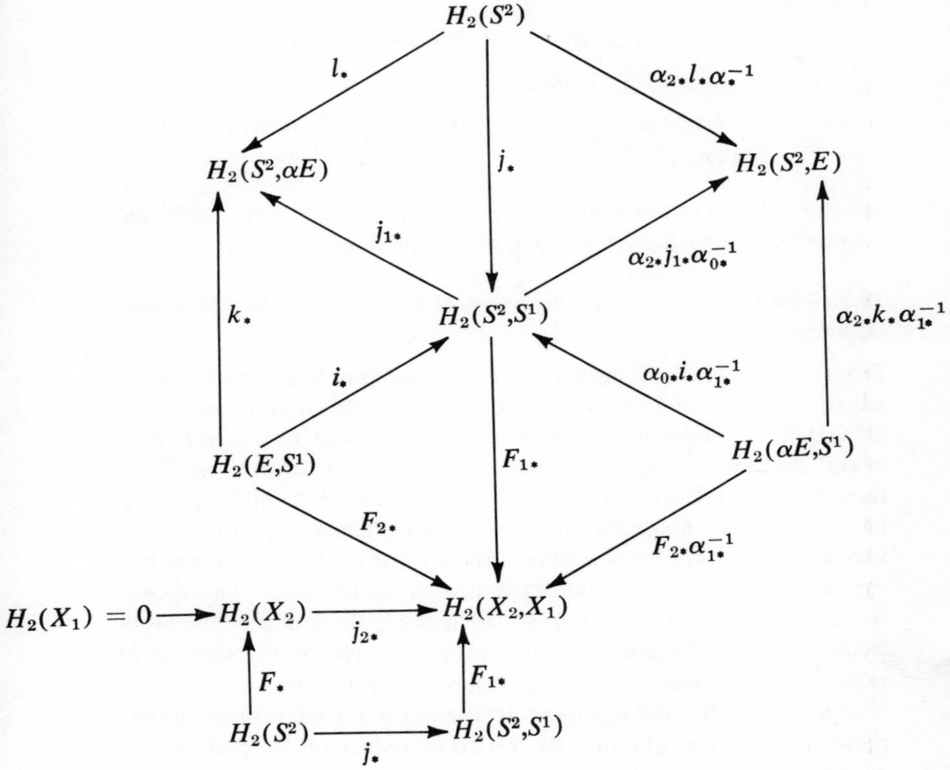

Since j_{2*} is a monomorphism, $F_* = 0$. In cohomology, an analogous argument shows that $F^* = 0$.

Definition: Adjoin E^3 to X_2 by the function $F: S^2 \longrightarrow X_2$. The space that arises is called the *lens space* $L(n, m)$.

The non-zero homology groups of X_2 were

$$H_1(X_2) \cong H^2(X_2) \cong M/nM,$$
$$H_2(X_2) \cong H^1(X_2) \cong M^{(n)}$$
$$H_0(X_2) \cong H^0(X_2) \cong M \text{ (in the non-augmented case).}$$

From 15.8 to 15.12, there follows, since $F_* = 0$,

18.8. THEOREM: *The non-zero homology and cohomology groups of $L(n, m)$ are*

$$H_0(L(n, m)) \cong M \qquad H^0(L(n, m)) \cong M$$
$$H_1(L(n, m)) \cong M/nM \qquad H^1(L(n, m)) \cong M^{(n)}$$
$$H_2(L(n, m)) \cong M^{(n)} \qquad H^2(L(n, m)) \cong M/nM$$
$$H_3(L(n, m)) \cong M \qquad H^3(L(n, m)) \cong M.$$

All the other homology and cohomology groups, in particular, the augmented groups $\tilde{H}_0(L(n, m))$ and $\tilde{H}^0(L(n, m))$, are zero.

18.9. THEOREM: *The lens space $L(n, m)$ is a compact 3-dimensional manifold.*

Proof: $L(n, m)$ is a spherical complex, and hence, a compact hausdorff space. To prove that it is a manifold, it is necessary to exhibit a euclidean neighborhood for each point of $L(n, m)$. It is trivial that each point of $\mathring{E}^3$ has euclidean neighborhood in $\mathring{E}^3$. It remains to be proved that each point of the image X_2 of E^2_+ under F has a euclidean neighborhood in $L(n, m)$. Pick an open set in E^3 which is formed by a small sphere around the north pole. There is a symmetrically placed neighborhood of the south pole. The identification requires that the cap at the north pole (the portion of S^2 within the neighborhood) be glued onto the cap at the south pole after the south polar cap has been turned through an angle of m/n. The result of the identification is obviously a euclidean neighborhood. For the investigation of the other points of X_2, pick $P_0, P_1, \cdots, P_{n-1}$, equally spaced points on the equator. Together with the north and south poles, the points P_i span a double pyramid that is homeomorphic to E^3.

It may be assumed that the angle NQS between faces meeting at the equatorial polygon is precisely $1/n$. The vertices P_i (i modulo n) yield a subdivision of the double pyramid in n tetrahedra $P_iP_{i+1}NS$. Designate the faces of these tetrahedra as follows: $P_iP_{i+1}N = a_i$, $P_iP_{i+1}S = b_i$, $P_iNS = c_i$, $P_{i+1}NS = d_i$. Here, d_i is identified with c_{i+1}. The lens space will now be reassembled from the tetrahedra in such a way as to describe neighborhoods of points of X_2. The function F requires that a_i be glued onto b_{i+m}. This is accomplished by placing the ith upon the $(i + m)$th tetrahedron. Since the angle at the equatorial polygon was chosen to be $1/n$ and m is relatively prime to n, the figure closes after all the tetrahedra have been joined. This process has formed a new double pyramid in which the

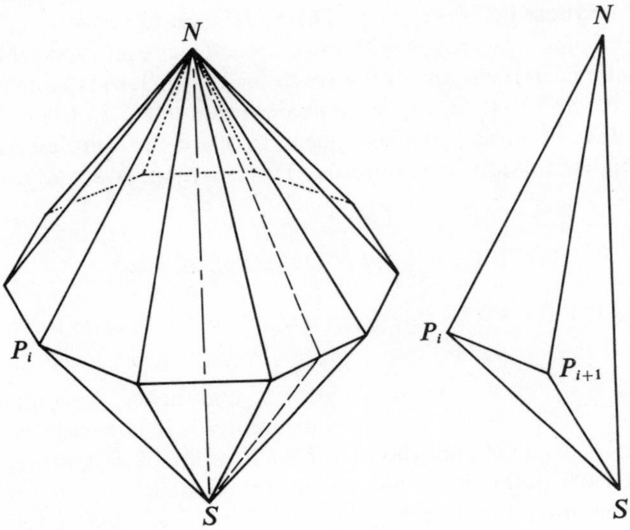

Figure 23

P_i form just the north and south poles. The c_i all lie around one pole, say the north, and the d_i all lie at the other. The points in the interior of a_i are inner points of the new double pyramid, and therefore have euclidean neighborhoods. Since the equatorial polygon can be turned, it has been proved that each point of X_2 has a euclidean neighborhood and that $L(n, m)$ is a manifold.

18.10. THEOREM: *The lens spaces $L(n, m)$ and $L(n, s)$ are homeomorphic when $sm \equiv 1$ modulo n.*

Proof: Continue the process of assembling the tetrahedra that was begun in the proof of 18.9 by glueing d_i to c_{i+1}. In this fashion, the ith tetrahedron has a face in common with the $(i + m)$th, this in turn has a face in common with the $(i + 2m)$th, etc. If s is the smallest solution of the congruence $sm \equiv 1$ modulo n, then d_i is precisely s steps away from c_{i+1} in this process. The identification of d_i with c_{i+1}, therefore, yields the lens space $L(n, s)$.

In a trivial fashion (by a reflection) it can be seen that $L(n, m)$ and $L(n, -m)$ are homeomorphic. In general, when n is fixed, certain of the $\phi(n)$ (Euler's ϕ function) possible lens spaces are homeomorphic. It was proved by E. J. Brody (*Annals of Mathe-*

matics, Vol. 71, (1963)) that the only homeomorphisms between lens spaces are those just described. (This is difficult to prove.)

From the homology groups with integer coefficients, it is possible to prove that n is an invariant. It is easy to see that $L(1, 0)$ is homeomorphic to S^3, and that $L(2, 1)$ is homeomorphic to P^3. In the case of larger values of n, new spaces appear that are the simplest examples of higher torsion in manifolds. Two such cases are $L(7, 1)$ and $L(7, 2)$.

19

Classification of Surfaces

Let X_1 and X_2 be two topological spaces adjoined as described just before 13.9 to form a space X. Let Y be the space obtained by attaching a ball E^n to X by means of the homeomorphism between the boundaries of E^n and E_1^n. The triad (Y, X_1, X_2) is exact and

$$X_1 \cup X_2 = Y$$
$$X_1 \cap X_2 = E^n \neq \varnothing.$$

Since $H_q(E^n) = 0$ (in the augmented theory), the Mayer-Vietoris sequence yields

19.1. $$H_q(Y) \cong H_q(X_1) \oplus H_q(X_2).$$

19.2. THEOREM: *Let $H_n(X_1) \cong H_n(X_2) = 0$. Then $H_n(X) = 0$. If, in addition, $M = R$ is an integral domain and $n \geq 2$, then*

$$B_{n-1}(X) = B_{n-1}(X_1) + B_{n-1}(X_2) + 1.$$

Proof: The sequence 15.8 becomes

$$0 \longrightarrow H_n(X) \longrightarrow H_n(Y) \longrightarrow H_{n-1}(S^{n-1}) \longrightarrow H_{n-1}(X) \longrightarrow H_{n-1}(Y) \longrightarrow 0.$$

By 19.1, $H(Y) = 0$, and consequently $H_n(X) = 0$. An application of $\otimes_R K$ to $H_{n-1}(S^{n-1}) \cong R$ yields $B_{n-1}(X) = B_{n-1}(Y) + 1$.

19.3. THEOREM: *Let Φ_h be the surface obtained by the successive adjoining of h projective planes. Let M be Z, the ring of integers. Then $H_2(\Phi_h) = 0$ and $B_1(\Phi_h) = h - 1$.*

Proof: Theorem 17.5 yields $H_2(\Phi_1) = H_2(P^2) = Z^{(2)} = 0$, and $H_1(\Phi_1) \cong Z/2Z$. This proves that $B_1(\Phi_1) = 0$. Now suppose that the theorem has been proved for Φ_{h-1}. Then it follows for Φ_h by letting $n = 2$, $X_1 = \Phi_{h-1}$, and $X_2 = \Phi_1$ in the previous theorem.

Remark: It was proved in 13.9 that $H_2(F_g) \cong Z$, and now it is known that $H_2(\Phi_h) = 0$. Hence, F_g and Φ_h are not homeomorphic or homotopic. The euler characteristic distinguished the F_g, and similarly, the Φ_h, from each other. Hence, no two F_g or Φ_h are homotopic, and they are not homeomorphic either.

Simplicial Complexes

Definitions: Let Δ be a standard simplex of sufficiently high dimension. A *simplicial complex* γ is a subset of the faces of Δ that statisfies the following condition: If $\Delta' \in \gamma$ and Δ'' is a face of Δ', then $\Delta'' \in \gamma$.

The union of the simplices of γ forms a subspace of Δ which will be denoted by $|\gamma|$. It is a closed subset of Δ, and therefore, a compact hausdorff space. The *dimension* of γ is the maximum of the dimensions of the simplices of γ.

A topological space is called *triangulable* if and only if there is a simplicial complex γ and a homeomorphism $t\colon |\gamma| \longrightarrow X$ of $|\gamma|$ on X. The homeomorphism t is called a *triangulation* of X.

A space pair (X, A) is called *triangulable* if and only if there is a simplicial complex γ and a subcomplex thereof γ_0 and a homeomorphism $t: (|\gamma|, |\gamma_0|) \longrightarrow (X, A)$. Triangulable spaces are compact and hausdorff. The simplicial complex for a triangulation is usually extremely difficult to picture since even simple figures such as the torus require a Δ of very high dimension. A *surface* F is a 2-dimensional manifold which is connected and triangulable. [It was proved by T. Rado in 1925 that each separable, compact, connected, 2-manifold is triangulable.]

19.4. THEOREM: *Let F be a surface, let γ be a simplicial complex, and let $t: |\gamma| \longrightarrow F$ be a triangulation. Then* $\dim \gamma = 2$.

Proof: Let $\Delta_1 \in \gamma$, and let x be an interior point of Δ_1. Then $t(x)$ has a euclidean 2-dimensional neighborhood, and x has a small neighborhood which can be embedded homeomorphically in R^2. By Corollary 14.9 (invariance of dimension) it follows that $\dim \Delta_1 \leq 2$. In case $\dim \Delta_1 < 2$, the homeomorphism t^{-1} would map a neighborhood of $t(x)$ on a neighborhood of x and thereby contradict 14.9. Therefore, each of the finite set of simplices in γ has dimension 2, and γ is 2-dimensional.

Since each point of F has a euclidean neighborhood, each 1-dimensional simplex in γ is face of exactly two 2-simplices of γ. For the same reason, each vertex of γ is vertex of precisely one cyclicly ordered sequence of 2-simplices in γ. These two conditions are obviously sufficient to guarantee that $|\gamma|$ is a surface.

If $t: |\gamma| \longrightarrow X$ is a triangulation of an n-dimensional manifold X, then a proof analogous to that of 19.4 will show that $\dim X = n$. The triangulability of compact connected 3-manifolds was established by E. Moise in 1951, but the question as to which compact connected manifolds of dimension at least four are triangulable is a difficult unsolved problem. Not even for 3-manifolds is there any approach to a solution to the homeomorphism problem, and no approach is known to reasonable restrictions of the problem. The subject is exceedingly complicated.

The object of this chapter is the classification of surfaces. It was already proved that no two of the surfaces F_g and Φ_h are homeomorphic. The next theorem will show that there are no other possible surfaces.

19.5. THEOREM: *Each surface is homomorphic to an F_g or a Φ_h.*

Proof: The proof of this theorem will occupy several pages and will consist of the reduction of F to a standard form. It will be assumed that F is already in the form of a simplicial complex.

Pick a triangle $\Delta_1 \in \gamma$ and a linear homeomorphism ϕ_1 sending Δ_1 into the plane R^2. Let k_1 be an edge of Δ_1; it is an edge of precisely one other triangle $\Delta_2 \in \gamma$. The restriction of ϕ_1 to k_1 can be extended to a linear homeomorphism $\phi_2 : \Delta_2 \longrightarrow R^2$ so that $\phi_1(\Delta_1)$ and $\phi_2(\Delta_2)$ are on opposite sides of $\phi_1(k_1) = \phi_2(k_2)$ and $\phi_1(\Delta_1) \cup \phi_2(\Delta_2)$ forms a convex quadrilateral Π.

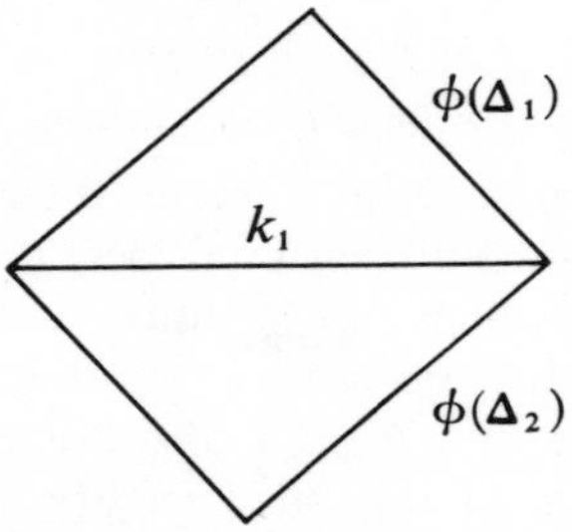

Figure 24

Suppose now that Π is a triangulated convex polygon in the plane which is piecewise linearly homeomorphic to a subset of the triangles of γ. Let s be a side of Π. Since Π is convex, it lies on one side or the other of the line determined by s. There is a triangle Δ_3 or γ and a linear homeomorphism ϕ_3 of Δ_3 on the triangle in Π whose side is s. Then s is image under ϕ_3 of an edge k in Δ_3. Now k is an edge of another triangle $\Delta_4 \in \gamma$. Suppose that Δ_4 is not yet mapped into the plane. Then the restriction of ϕ_3 to k can be continued to a linear homeomorphism $\phi_4 : \Delta_4 \longrightarrow R^2$ so that Π and $\phi_4(\Delta_4)$ lie on opposite sides of s and $\Pi \cap \phi_4(\Delta_4)$ is still a convex polygon. Continue this process until all the triangles of γ have been mapped. This is possible since $|\gamma|$ is connected. The final result is a convex polygon Π whose edges are pairwise linearly homeomorphic. If these pairs are identified, the resulting figure is homeomorphic to F.

Choose a sense of traversing the boundary of the polygon Π. Let s be a side of Π. Orient s independently of the sense of traversal; give it the new name a if the orientation and direction of traversal coincide, and name it a^{-1} if the orientation and direction of transversal are opposite. The side homeomorphic to s will have an orientation determined by that of s and by the homeomorphism. It will also be named a or a^{-1} depending on whether or not its orientation coincides with the direction of traversal.

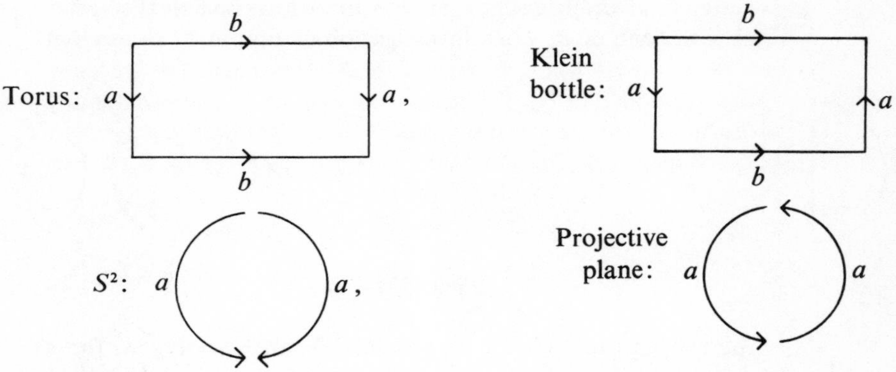

Figure 25

Each surface is now described by the non-commutative product of its edges in the order of the sense of traversal. For instance, by $aba^{-1}b^{-1}$ in the case of the torus, and aa^{-1} for S^2. The triangulation of S^2 by a tetrahedron shows that S^2 also can be described by $aa^{-1}bb^{-1}cc^{-1}$. Let A and B be two such products (symbols for a surface). The equivalence $A \sim B$ shall mean that A and B are symbols for homeomorphic surfaces.

The following transformation rules lead from one symbol for a surface to an equivalent one:

(1) Cyclic permutations of the sides of Π, renaming of the sides, reorienting the sides (replacing x by x^{-1}).

(2) Let sub-products in one of the symbols be designated by Roman capitals. Let F correspond to AB where A and B do not overlap.

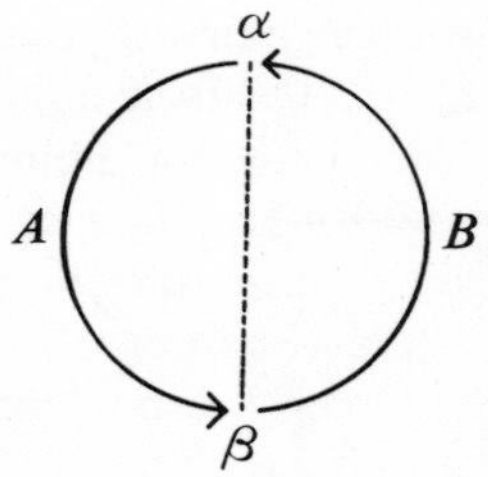

Figure 26

Suppose that the glueing prescription for A indicates that the vertex α at the beginning of A is to be attached to the vertex β at the end of B. Then investigation of Figure 27 shows that

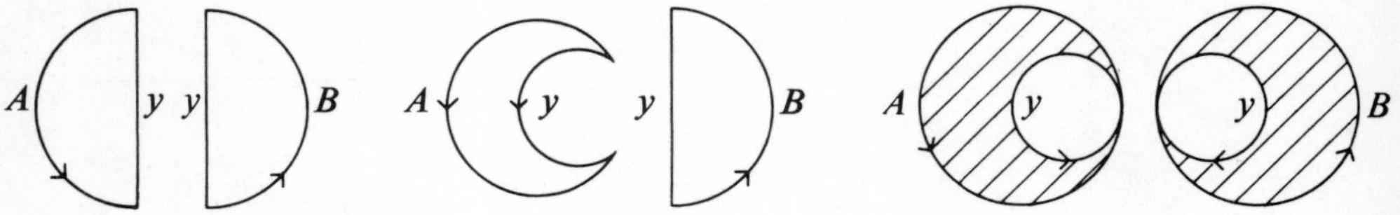

Figure 27

the polygon for AB can be obtained from the polygons for A and B by the removal of a circular disc and a glueing process. This will be used only in the cases

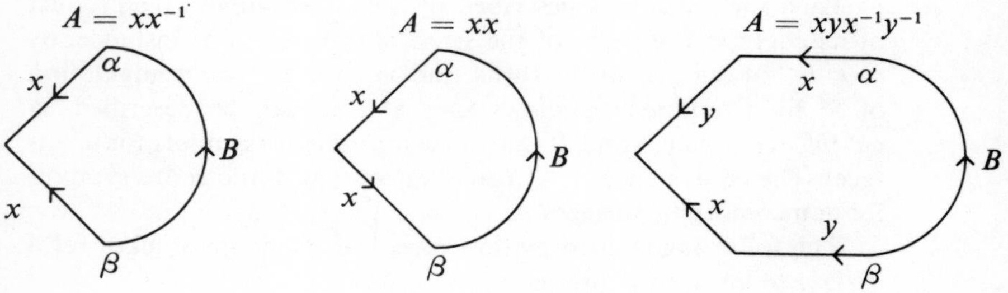

Figure 28

(3) The relation $xx^{-1}B \sim B$ follows from (2) as can be seen from the fact that xx^{-1} represents a sphere S^2. It can be proved directly as follows:

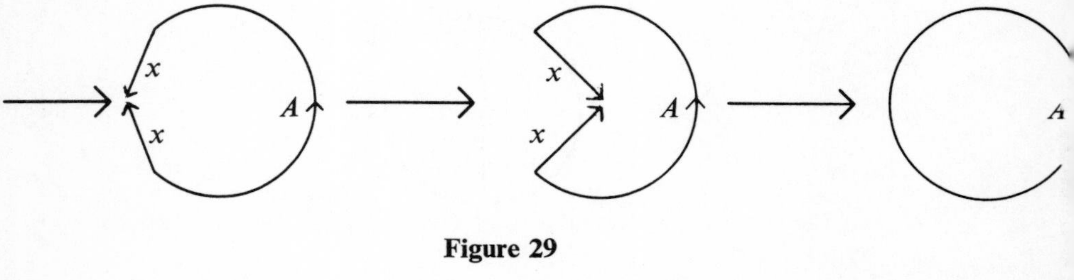

Figure 29

Since the formula is general, the suface symbol 1 is defined to be the one belonging to S^2.

(4) $$AxBCx^{-1} \sim AxCBx^{-1}.$$

Proof: After making the cut and identification indicated in Figure 30, rename y to be x.

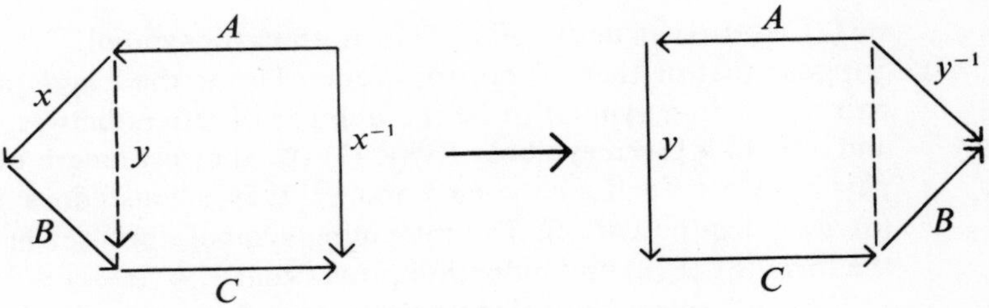

Figure 30

This equivalence shows that the letters between x and x^{-1} may be permuted cyclically.

(5) $$xAxB \sim xxAB^{-1}$$

Proof: After making the cut and identification indicated in Figure 31, rename y to be x.

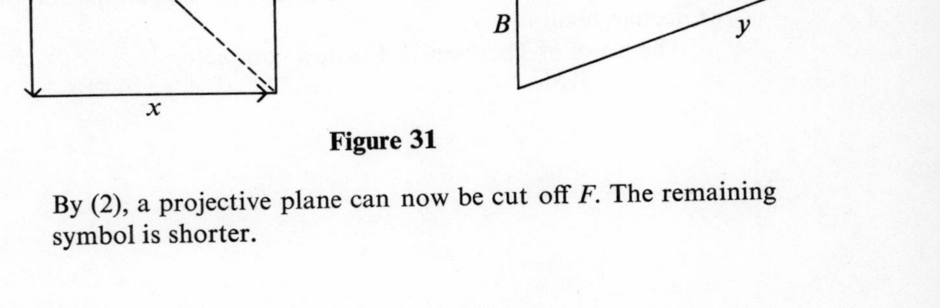

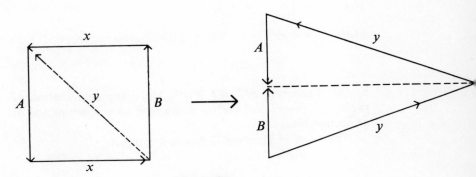

Figure 31

By (2), a projective plane can now be cut off F. The remaining symbol is shorter.

(6) Suppose that there are in F two pairs x, x^{-1}, y, y^{-1} which separate each other. Then, by cyclic permutation, it follows that

$$\begin{aligned} F = AxByCx^{-1}Dy^{-1} &\sim AxByDCx^{-1}y^{-1} \\ &\sim AxDCByx^{-1}y^{-1} \sim y^{-1}AxDCByx^{-1} \\ &\sim y^{-1}DCBAxyx^{-1} \sim DCBAxyx^{-1}y^{-1}. \end{aligned}$$

By (2), a torus can be cut off F. This shortens the symbol.

(7) Suppose that neither (5) nor (6) occurs. Define the *length* of a pair x, x^{-1} in a symbol to be the number of letters between x and x^{-1}. In a given symbol of type (7) the shortest length of a pair is 0 since if y lies between x and x^{-1} then y^{-1} also does. By (3), xx^{-1} can be cut off. The remaining symbol does not have the form (5) or (6) and induction proves that $F \sim 1$.

The first seven equivalences prove that F is either a sphere or assembled from tori and projective planes. If only tori occur, F is a surface F_g of genus g.

(8) If a projective plane P^2 can be cut off F, then F can be assembled from nothing but projective planes. It is then a Φ_h.

Proof:

$$\begin{aligned} P^2T = xxyzy^{-1}z^{-1} &\sim z^{-1}xxyzy^{-1} \\ &\sim z^{-1}xyxzy^{-1} \sim xyxzy^{-1}z^{-1} \\ &\sim xxy^{-1}zy^{-1}z^{-1} \sim xxyzyz^{-1} \\ &\sim P^2K \end{aligned}$$

where $K = yzyz^{-1}$ is the Klein bottle. Furthermore,

$$K = yzyz^{-1} \sim yyzz = P^2P^2$$

by (5). Therefore, $P^2T \sim P^2P^2P^2$. Thus, any combination of projective planes and tori is equivalent to a combination of projective planes.

The proof of Theorem 19.5 is now complete.

20

Singular Cup Products

The homology theory employed in this chapter is the non-augmented singular theory.

Let M, N, and P be R-modules and let $M \times N \longrightarrow P$ be a bilinear function written $m \circ n$. As before, $S_p(X, A)$ is the free R-module generated by the p-simplices of (X, A); that is, those p-simplices whose carriers do not lie entirely in A. A cochain $c^p \in S^p(X, A)$ is a homomorphism

$$c^p : S_p(X, A) \longrightarrow M,$$

and is thus completely determined by its values on the p-simplices of (X, A). In order to permit c^p to be defined on arbitrary simplices of X, the convention $c^p(\sigma) = 0$ can be made for simplices σ whose carrier lies in A. This convention will be employed from now on.

In the case $p < 0$, $S^p(X, A; M) = 0$. Furthermore, by 11.5,

$$(\delta(c^p))\sigma = c^p\partial\sigma$$

where $\sigma \in S_{p+1}(X, A)$; hence, by 3.9

$$(\delta c^p)\sigma = c^p\sigma\partial(\Delta^{\dagger}_{p+1}).$$

This means that $\delta c^p = 0$, when $p < 0$.

Now let $\sigma \in S_{p+q}(X, A)$, and let

$$(d_0, d_1, \cdots, d_p)\colon \Delta_p \longrightarrow \Delta_{p+q}$$
$$(d_p, d_{p+1}, \cdots, d_{p+q})\colon \Delta_q \longrightarrow \Delta_{p+q}$$

be the functions defined just prior to Theorem 3.2. Here, in contrast to Chapter 3, the ranges of the latter two functions have been restricted to Δ_{p+q}.

Definition: To each ordered pair (c^p, c^q) where $c^p \in S^p(X, A; M)$ and $d^q \in S^q(X, A; N)$ is associated a $p + q$-cochain

$$c^p \smile d^q \in S^{p+q}(X, A, P)$$

by means of

$$(c^p \smile d^q)(\sigma) = \begin{cases} 0, \text{ when } p \text{ or } q \text{ is negative} \\ c^p(\sigma(d_0, \cdots, d_p) \circ d^q(\sigma(d_p, \cdots, d_{p+q})), \\ \text{when } p \text{ and } q \text{ are non-negative.} \end{cases}$$

It is clear that $c^p \smile d^q$ is zero whenever one of the factors is, and that the operation $\smile$ defines a bilinear function

$$S^p(X, A; M) \times S^q(X, A; N) \longrightarrow S^{p+q}(X, A; P).$$

In other words, the cup product is distributive and R-homogeneous.

20.1. THEOREM: $\delta(c^p \smile d^q) = \delta c^p \smile d^q + (-1)^p c^p \smile \delta d^q$, *where* $c^p \in S^p(X, A; M)$ *and* $d^q \in S^q(X, A; N)$.

Proof:

(1) If $p < 0$, then $c^p \smile d^q = 0$, $\delta c^p = 0$, and $c^p = 0$. A similar statement holds for $q < 0$.

(2) Let p and q be non-negative. It is sufficient to investigate the action of the terms of the equation on an arbitrary $p + q + 1$ simplex as follows:

$$\begin{aligned}\delta(c^p \smile d^q)(\tau) &= (c^p \smile d^q)(\tau\partial(\Delta^{\dagger}_{p+q+1}))\\ &= \sum_{j=0}^{p+q+1} (-1)^j (c^p \smile d^q)(\tau(d_0, \cdots, \hat{d}_j, \cdots, d_{p+q+1}))\\ &= \sum_{j=0}^{p+q+1} (-1)^j c^p(\tau(d_0, \cdots, \hat{d}_j, \cdots, d_{p+q+1})(d_0, \cdots, d_p)) \circ\\ &\qquad d^q(\tau(d_0, \cdots, \hat{d}_j, \cdots, d_{p+q+1})(d_p, \cdots, d_{p+q})).\end{aligned}$$

In this last, lengthy sum it is easy to see that, when $j \leq p$,

$$(d_0, \cdots, \hat{d}_j, \cdots, d_{p+q+1})(d_0, \cdots, d_p) = (d_0, \cdots, \hat{d}_j, \cdots, d_{p+1})$$
$$(d_0, \cdots, \hat{d}_j, \cdots, d_{p+q+1})(d_p, \cdots, d_{p+q}) = (d_{p+1}, \cdots, d_{p+q+1}),$$

and that, when $j > P$,

$$(d_0, \cdots, \hat{d}_j, \cdots, d_{p+q+1})(d_0, \cdots, d_p) = (d_0, \cdots, d_p)$$
$$(d_0, \cdots, \hat{d}_j, \cdots, d_{p+q+1})(d_p, \cdots, d_{p+q}) = (d_p, \cdots, \hat{d}_j, \cdots, d_{p+q+1}).$$

Hence,

20.2.

$$\begin{aligned}&\delta(c^p \smile d^q)(\tau) =\\ &\quad\left[\sum_{j=0}^{p} (-1)^j c^p(\tau(d_0, \cdots, \hat{d}_j, \cdots, d_{p+1})\right] \circ d^q(\tau(d_{p+1}, \cdots, d_{p+q+1})) +\\ &\qquad c^p(\tau(d_0, \cdots, d_p)) \circ \left[\sum_{j=p+1}^{p+q+1} (-1)^j d^q(\tau(d_p, \cdots, d_j, \cdots, d_{p+q+1}))\right].\end{aligned}$$

On the other hand,

20.3.

$$\begin{aligned}(\delta c^p \smile d^q)(\tau) &= (\delta c^p)(\tau(d_0, \cdots, d_{p+1}) \circ d^q(\tau(d_{p+1}, \cdots, d_{p+q+1}))\\ &= c^p(\tau(d_0, \cdots, d_{p+1})\partial(\Delta^{\dagger}_{p+1})) \circ d^q(\tau(d_{p+1}, \cdots, d_{p+q+1}))\\ &= \left[\sum_{j=0}^{p+1} (-1)^j c^p(\tau(d_0, \cdots, d_{p+1})(d_0, \cdots, \hat{d}_j, \cdots, d_{p+1}))\right] \circ\\ &\quad d^q(\tau(d_{p+1}, \cdots, d_{p+q+1}))\\ &= \left[\sum_{j=0}^{p+1} (-1)^j c^p(\tau(d_0, \cdots, \hat{d}_j, \cdots, d_{p+1}))\right] \circ\\ &\quad d^q(\tau(d_{p+1}, \cdots, d_{p+q+1})),\end{aligned}$$

and

20.4.

$$
\begin{aligned}
(-1)^p(c^p \smile \delta d^q)(\tau)
&= (-1)^p c^p(\tau(d_0, \cdots, d_p)) \circ (\delta d^q)(\tau(d_p, \cdots, d_{p+q+1})) \\
&= (-1)^p c^p(\tau(d_0, \cdots, d_p)) \circ d^q(\tau(d_p, \cdots, d_{p+q+1})\partial(\Delta^{\dagger}_{q+1}) \\
&= (-1)^p c^p(\tau(d_0, \cdots, d_p)) \circ \\
&\qquad \left[\sum_{j=0}^{q+1} (-1)^j d^q(\tau(d_p, \cdots, d_{p+q+1})(d_0, \cdots, \hat{d}_j, \cdots, d_{q+1}))\right] \\
&= (-1)^p c^p(\tau(d_0, \cdots, d_p)) \circ \\
&\qquad \left[\sum_{j=0}^{q+1} (-1)^j d^q(\tau(d_p, \cdots, \hat{d}_{p+j}, \cdots, d_{p+q+1}))\right] \\
&= c^p(\tau(d_0, \cdots, d_p)) \circ \\
&\qquad \left[\sum_{j=p}^{p+q+1} (-1)^j d^q(\tau(d_p, \cdots, \hat{d}_j, \cdots, d_{p+q+1}))\right].
\end{aligned}
$$

From 20.3 and 20.4, the sums on the right hand side of 20.2 can be evaluated to yield

$$
\begin{aligned}
\delta(c^p \smile d^q) = (\delta c^p \smile d^q)(\tau) - (-1)^{p+1} c^p(\tau(d_0, \cdots, d_p)) \circ \\
d^q(\tau(d_{p+1}, \cdots, d_{p+q+1})) \\
+ (-1)^p(c^p \smile \delta d^q)(\tau) - (-1)^p c^p(\tau(d_0, \cdots, d_p)) \circ \\
d^q(\tau(d_{p+1}, \cdots, d_{p+q+1})).
\end{aligned}
$$

From here, the theorem follows immediately. Notice that the theorem is a kind of differentiation of products.

20.5. Corollary: *If* $\delta c^p = \delta d^q = 0$, *then* $\delta(c^p \smile d^q) = 0$; *in other words,*

$$\textit{cocycle} \smile \textit{cocycle} = \textit{cocycle}.$$

20.6. Corollary: *If* $\delta d^q = 0$, *then* $\delta c^p \smile d^q = \delta(c^p \smile d^q)$; *in other words,*

$$\textit{coboundary} \smile \textit{cocycle} = \textit{coboundary}.$$

20.7. Corollary: *If* $\delta c^p = 0$, *then* $c^p \smile \delta d^q = (-1)^p \delta(c^p \smile d^q)$; *in other words,*

$$\textit{cocycle} \smile \textit{coboundary} = \textit{coboundary}.$$

Definition: Let $\alpha \in H^p(X, A; M)$ and $\beta \in H^q(X, A; N)$, and let c^p and d^q be elements of α and β respectively. Then $c^p \smile d^q$ is a cocycle and in some element of $H^{p+q}(X, A; P)$. If the choice of c^p or d^q is changed by adding a coboundary to either, the product $c^p \smile d^q$ is altered only by the addition of a coboundary to $c^p \smile d^q$. The element of $H^{p+q}(X, A, P)$ that contains $c^p \smile d^q$, thus defined by the choice of α and β, is called the *cup product* of α and β, and is denoted by $\alpha \smile \beta$. The cup product defines a bilinear function

$$H^p(X, A; M) \times H^q(X, A; N) \longrightarrow H^{p+q}(X, A; P).$$

Definition: M is a *graded module* if and only if it is the direct sum of a collection of modules indexed by the integers. If $M = \sum_q^{\oplus} M_q$ is a graded module, an element m of M is called *homogenous of degree* q if and only if there is a q for which $m \in M_q$. The zero element does not have a well-determined degree. Let R and S and M be graded modules and let $f: R \times S \longrightarrow M$ be bilinear. For f to be called a *bilinear function of graded modules,* it will be required of f that if $a \in R$ and $b \in S$ are homogenous of degrees r and s respectively, then $f(a, b)$ shall be homogenous of degree $r + s$.

An example of a graded module is the set of polynomials in n variables with coefficients in some ring. Here the direct summands indexed by negative integers are zero, and the word, "degree," has its usual meaning.

The modules $H_*(X, A) = \sum_q^{\oplus} H_q(X, A)$ and $H^*(X, A) = \sum_q^{\oplus} H^q(X, A)$ are graded modules. The cup product defines a bilinear function

$$H^*(X, A; M) \times H^*(X, A; N) \longrightarrow H^*(X, A; P).$$

The proof of many theorems in homology and cohomology theory becomes simpler if the graded modules $H_*(X, A)$ or $H^*(X, A)$ are employed, rather than the individual homology or cohomology groups.

Definition: A *homomorphism* $f: M \longrightarrow N$ *of graded modules* is a homomorphism carrying homogenous elements into homogenous elements in such a way that $\deg m - \deg f(m)$ is a constant on the set of homogenous elements m of M. This constant is called the *degree* of the homomorphism.

If $f: (X, A) \longrightarrow (Y, B)$ is continuous, then both f^* and f_* have degree zero since $f^*: H^q \longrightarrow H^q$ and $f_*: H_q \longrightarrow H_q$. On the other hand, ∂_* has degree -1 and δ^* has degree $+1$.

In terms of H_* and H^*, the usual exact sequences take the form of exact triangles (in which the image under each homomorphism is the kernel of the next):

$$\begin{array}{ccc} H_*(X, A) & \xrightarrow{\partial_*} & H_*(A) \\ {\scriptstyle j_*}\nwarrow & & \swarrow{\scriptstyle i_*} \\ & H_*(X) & \end{array} \qquad \begin{array}{ccc} H^*(X, A) & \xleftarrow{\delta^*} & H^*(A) \\ {\scriptstyle j^*}\searrow & & \nearrow{\scriptstyle i^*} \\ & H^*(X). & \end{array}$$

In contrast to the case of commutative triangles, all the homomorphisms here yield the same sense of rotation in the triangle.

The Mayer-Vietoris sequences also yield exact triangles:

$$\begin{array}{ccc} H_*(X) & \xrightarrow{\Delta} & H_*(A) \\ {\scriptstyle \phi}\nwarrow & & \swarrow{\scriptstyle \psi} \\ & H_*(X_1) + H_*(X_2) & \end{array} \qquad \begin{array}{ccc} H^*(X) & \xleftarrow{\Delta} & H^*(A) \\ \searrow & & \nearrow{\scriptstyle \phi} \\ & H^*(X_1) + H^*(X_2). & \end{array}$$

Let M, N, Q be modules for which bilinear products are defined between M and N, N and Q, $M \times N$ and Q, and M and $N \times Q$ such that $(m \circ n) \circ q = m \circ (n \circ q)$. Let c^p, d^q, e^r be cochains from the cochain complexes $S^p(X, A; M)$, $S^q(X, A; N)$ and $S^r(X, A; Q)$, respectively.

20.8. THEOREM:

$$(c^p \smile d^q) \smile e^r = c^p \smile (d^q \smile e^r).$$

Proof: Let σ be a $(p + q + r)$-simplex. Then

$$\begin{aligned} [(c^p \smile d^q) \smile e^r]\sigma &= (c^p \smile d^q)(\sigma(d_0, \cdots, d_{p+q})) \circ e^r(\sigma(d_{p+q}, \cdots, d_{p+q+r})) \\ &= c^p(\sigma(d_0, \cdots, d_{p+q})(d_0, \cdots, d_p)) \circ \\ &\qquad d^q(\sigma(d_0, \cdots, d_{p+q})) \circ e^r(\sigma(d_{p+q}, \cdots, d_{p+q+r})) \\ &= c^p(\sigma(d_0, \cdots, d_p)) \circ d^q(\sigma(d_p, \cdots, d_{p+q})) \circ e^r(\sigma(d_{p+q}, \cdots, d_{p+q+r})). \end{aligned}$$

A similar computation yields the same result for $[c^p \smile (d^q \smile e^r)]\sigma$.

As a consequence of 13.8, it is possible to write $c^p \smile d^q \smile e^r$.

20.9. Corollary (The Associative Law for Cup Products): *If* $\alpha \in H^*(X, A; M)$, $\beta \in H^*(X, A; N)$ *and* $\gamma \in H^*(X, A; Q)$, *then*

$$(\alpha \smile \beta) \smile \gamma = \alpha \smile (\beta \smile \gamma).$$

Suppose that the product $m \circ n$ is an element of N (has been defined by a bilinear function $M \times N \longrightarrow N$), and suppose that M has a unit element which is written 1. Let c^0 be the cochain which has the value 1 on each 0-simplex (point).

20.10. THEOREM: *The cochain* c^0 *is a cocycle; and, for each* $d^q \in S^q(X, A; N)$,

$$c^0 \smile d^q = d^q.$$

Proof: Let σ be a 1-simplex, then

$$\begin{aligned} \delta c^0(\sigma) &= c^0(\sigma\partial(\Delta_1^\dagger)) = c^0(\sigma(d_1) - (d_0))) \\ &= c^0(\sigma(d_1)) - c^0(\sigma(d_0)) = 1 - 1 = 0. \end{aligned}$$

Therefore, c^0 is a cocycle. For each q-simplex σ,

$$\begin{aligned} (c^0 \smile d^q)\sigma &= c^0(\sigma(d_0)) \circ d^q(\sigma(d_0, \cdots, d_q) \\ &= 1 \circ d^q(\sigma) = d^q(\sigma). \end{aligned}$$

Therefore, c^0 *is a left-hand identity.*

20.11. Corollary: *The cohomology class containing* c^0 *is a left-hand identity for the cup product of elements of* $H^*(X, A; M)$ *and* $H^*(X, A; N)$.

It is clear that analogous statements can be made when bilinear function $M \times N \longrightarrow M$ has a unit element in N. The interesting case is the one in which $M = N$, M is an associative ring with a unit, and cup products of pairs of elements of $H^*(X, A; M)$ are being considered. Then the results of 13.9, 13.10, and 13.11 yield

20.12. THEOREM: *If* M *is an associative ring with a unit element, then* $H^*(X, A; M)$ *is a graded, associative ring with a two-sided unit element.*

Definition: The ring in Theorem 13.2 is called the *homology ring* of the pair (X, A). M is often taken to be the ring of integers.

Examples:

(1) $H^*(S^n; M)$ is the direct sum of two homogenous submodules: $H^0(S^n)$ and $H^n(S^n)$ both of which are isomorphic to M; that is, $H^*(S^n; M)$ has the form $Re_0 \oplus Re_n$. Here, e_0 is clearly the unit element of $H^*(S^n, M)$. Only the product $e_n \smile e_n$ remains to be investigated; but this product has degree $2n$ and is therefore zero. The module $H^*(S^n, M)$ is the "module of dual numbers" of Stedy.

(2) The torus and the figure obtained by attaching two disjoint circles to a 2-sphere have isomorphic cohomology groups in all dimensions, but their cohomology rings (with integer coefficients) do not have the same multiplicative structure in dimension 1. The computations here are left to the reader.

Let $f: (X, A) \longrightarrow (Y, B)$ be continuous. By definition,

$$f^{\#}(c^p\sigma) = c^p(f(\sigma)).$$

It follows immediately that

20.13. $$f^{\#}(c^p \smile d^q) = (f^{\#}c^p) \smile (f^{\#}d^q)$$

where c^p and d^q are cocycles, and that

20.14. $$f^*(\alpha \smile \beta) = f^*(\alpha) \smile f^*(\beta)$$

where α and β are in cohomology groups of (X, A) with the given coefficients. *Thus, f^* commutes with cup multiplication.*

In case a cohomology ring of (X, A) is under consideration, Equation 20.14 shows that f^* is not just a module homomorphism. It is also a ring homomorphism of degree zero.

21

The Singular Cap Product

This chapter defines a product of a chain by a cochain which yields a chain, and the corresponding product of the homology by the cohomology group is investigated.

Definition: Let $M \times N \longrightarrow P$ be a bilinear function denoted by the product $m \circ n$. Let $c^p \in S^p(X, M)$ and let $c_q \in S_q(X, N)$. Here $c_q = 0$ if $q < 0$, and, if $q \geq 0$, $c_q = \sum_j b_j \sigma_j$, where $b_j \in N$ and σ_j is a q-simplex. The *cap product* $c^p \frown c_q$ associates with the pair (c^p, c_q) a $(p - q)$-chain of $S_{p-q}(X, P)$ by means of

$$c^p \frown c_q = 0 \qquad \text{when } q < 0 \text{ or } p < 0$$

$$c^p \frown c_q = \textstyle\sum_j c^p \frown b_j \sigma_j \qquad \text{when } q \geq 0 \text{ and } c_q = \sum_j b_j \sigma_j$$

and

21.1.

$$c^p \frown b\sigma = \begin{cases} 0 & \text{if } p > q \geq 0 \\ [c^p(\sigma(d_{q-p}, \cdots, d_q)) \circ b](\sigma(d_0, \cdots, d_{p-q})) & \text{if } q \geq p \geq 0, \end{cases}$$

where the affine simplices involved have images in Δ_q. Then $c^p \frown c_q \in S_{q-p}(X)$, and the cap product yields a bilinear function

$$S^p(X, M) \times S_q(X, N) \longrightarrow S_{q-p}(X, P).$$

21.2. THEOREM: *If c^p is a p-cochain and c_q is a q-chain, then*

$$\partial(c^p \frown c_q) = (-1)^{q-p}(\delta c^p \frown c_q) + (c^p \frown \partial c_q).$$

Proof: Before beginning the calculations, notice that the dimensions of the chains on the right are correct.

(1) If $q < 0$, all the terms are zero.

(2) If $p > q \geq 0$, the left-hand side is zero by the definition of the cap product. The right-hand side is also zero since $p + 1 > q$ and $p > q - 1$.

(3) If $p < 0$, then $c^p = 0$ and $\delta c^p = 0$.

(4) Let $0 \leq p \leq q$. It is sufficient to consider the case $c_q = b\sigma$, where σ is a q-simplex and $b \in N$. In this case,

$$\partial(c^p \frown b\sigma) = (c^p \frown b\sigma)\partial(\Delta^{\dagger}_{q-p}),$$

where $\partial(\Delta^{\dagger}_{q-p}) = 0$ when $q = p$ since the non-augmented homology theory is being used.

In the case $q = p$ the left-hand side of 14.2 is zero, and it must be proved that

$$0 = (\delta c^p \frown b\sigma) + (c^p \frown b\partial c) \text{ for } \sigma = \sigma_p.$$

Fortunately, both summands are zero since $p + 1 > p$ and $p > p - 1$.

Finally, when $p < q$, it is possible to calculate with the usual boundary formulas to obtain

21.3.
$$\begin{aligned}\partial(c^p \frown b\sigma) &= (c^p \frown b\sigma)\partial(\Delta^\dagger_{p-q}) \\ &= \sum_{i=0}^{q-p}(-1)^i[c^p\sigma(d_{q-p}, \cdots, d_q) \circ b]\cdot \\ &\qquad (\sigma(d_0, \cdots, d_{q-p}))(d_0, \cdots, \hat{d}_i, \cdots, d_{q-p}) \\ &= \sum_{i=0}^{q-p}(-1)^i[c^p\sigma(d_{q-p}, \cdots, d_q) \circ b]\cdot \\ &\qquad (\sigma(d_0, \cdots, \hat{d}_i, \cdots, d_{q-p})).\end{aligned}$$

On the other hand, the last term of 21.2 is

$$(c^p \frown \partial b\sigma) = \sum_{i=0}^{q}(-1)^i(c^p \frown b\sigma(d_0, \cdots, \hat{d}_i, \cdots d_q))$$

where each term in the sum on the right is the cap product of c^p with a $(q - 1)$-chain; hence, by 21.1, this sum becomes

$$\sum_{i=0}^{q}(-1)^i[c^p\sigma(d_0, \cdots, \hat{d}_i, \cdots, d_q)(d_{q-1-p}, \cdots, d_{q-1}) \circ b]\cdot$$
$$(\sigma(d_0, \cdots, \hat{d}_i, \cdots, d_q))(d_0, \cdots, d_{q-1-p}).$$

Here,

$$(d_0, \cdots, \hat{d}_i, \cdots, d_q)(d_{q-1-p}, \cdots, d_{q-2}) = \begin{cases}(d_{q-p}, \cdots, d_q) & \text{when } i < q - p \\ (d_{q-1-p}, \cdots, \hat{d}_i, \cdots, d_q) & \text{when } i \geq q - p\end{cases}$$

and

$$(d_0, \cdots, \hat{d}_i, \cdots, d_q)(d_0, \cdots, d_{q-1-p}) = \begin{cases}(d_0, \cdots, \hat{d}_i, \cdots, d_{q-p}) & \text{when } i < q - p \\ (d_0, \cdots, d_{q-1-p}) & \text{when } i \geq q - p.\end{cases}$$

Thus,

21.4.
$$\begin{aligned}(c^p \frown \partial b\sigma) &= \sum_{i=0}^{q-p-1}(-1)^i[c^p\sigma(d_{q-p}, \cdots, d_q) \circ b]\cdot \\ &\qquad (\sigma(d_0, \cdots, \hat{d}_i, \cdots, d_{q-p})) \\ &\quad + \sum_{i=q-p}^{q}(-1)^i[c^p\sigma(d_{q-1-p}, \cdots, \hat{d}_i, \cdots, d_q) \circ b](\sigma(d_0, \cdots, d_{q-1-p})).\end{aligned}$$

The first term on the right-hand side of Equation 21.2 is

$$(-1)^{q-p}(\delta c^p \frown d_q) = (-1)^{q-p}[\delta c^p(\sigma(d_{q-1-p}, \cdots, d_q) \circ b]\cdot (\sigma(d_0, \cdots d_{q-1-p}))$$

$$= (-1)^{q-p}[c^p(\sigma)d_{q-1-p}, \cdots, d_q))\partial(\Delta^{\dagger}_{p+1}) \circ b](\sigma(d_0, \cdots, d_{q-1-p})$$

$$= \sum_{i=0}^{p+1} [(-1)^{q-p+i}c^p(\sigma(d_{q-1-p}, \cdots, d_q)(d_0, \cdots, \hat{d}_i, \cdots, d_{p+1})) \circ b]\cdot (\sigma(d_0, \cdots, d_{q-1-p}))$$

$$= \sum_{i=0}^{p+1} [(-1)^{q-p+i}c^p(\sigma(d_{q-1-p}, \cdots, \hat{d}_{q-1-p+i}, \cdots, d_q)) \circ b]\cdot (\sigma(d_0, \cdots, d_{q-1-p})).$$

Replacement of $q - 1 - p + i$ by i in this last sum yields

21.5. $$(-1)^{q-p}(\delta c^p \frown b\sigma) = [(-1)^{q-p}c^p\sigma(d_{q-p}, \cdots, d_q) \circ b]\cdot (\sigma(d_0, \cdots, d_{q-1-p})) + \sum_{i=q-p}^{q} [(-1)^{i-1}c^p(\sigma(d_{q-1-p}, \cdots, \hat{d}_i, \cdots, d_q) \circ b]\cdot (\sigma(d_0, \cdots, d_{q-1-p})).$$

In an addition of Equations 21.4 and 21.5, the second terms on the right-hand side vanish and the remaining terms are those on the right-hand side of 21.3. This establishes the theorem.

21.6. Corollary: *The cap product of a cocycle and a cycle is a cycle, the cap product of a coboundary and a cycle is a bounding cycle, and the cap product of a cocycle and a boundary is a boundary.*

21.7. Corollary: *The cap product induces a bilinear function*

$$H^p(X; M) \times H_q(X, N) \longrightarrow H_{q-p}(X, P)$$

which is also termed the cap product, and it induces a bilinear function

$$H^*(X, M) \times H_*(X, N) \longrightarrow H_*(X, P)$$

whereby the homogenous elements of degrees (p, q) are sent into homogenous elements of degree $q - p$.

As in the case of cup products, let M, N, and Q be modules for which $(m \circ n) \circ q = m \circ (n \circ q)$ and all the products of this form are defined and lie in a module P.

21.8. THEOREM: *Let* $c^r \in S^r(X, M)$, $d^p \in S^p(X, N)$, *and* $e_q \in S_q(X, Q)$ *then*

$$c^r \frown (d^p \frown e_q) = (c^r \smile d^p) \frown e_q.$$

Proof:

(1) If one of r, p, or q is negative, then both sides of Equation 21.8 are zero. Hence, all three can be assumed to be non-negative.

(2) If $p + r > q$, then $r > q - p$, and both sides of 14.8 are zero. Hence, it can be assumed that $p + r \leq q$.

(3) If r, p, and q are non-negative and $p + r \leq q$, then it is sufficient to prove 14.8 in the special case that e_q is the chain $b\sigma$, where $b \in Q$ and σ is a simplex in X. Then

$$\begin{aligned} &c^r \frown (d^p \frown b\sigma) \\ &\quad = c^r \frown \{[d^p(\sigma(d_{q-p}, \cdots, d_q)) \circ b](\sigma(d_0, \cdots, d_{q-p}))\} \\ &\quad = [c^r(\sigma(d_0, \cdots, d_{q-p})(d_{q-p-r}, \cdots, d_{q-p})) \circ d^p(\sigma(d_{q-p}, \cdots, d_q) \circ b] \cdot \\ &\qquad\qquad (\sigma(d_0, \cdots, d_{q-p})(d_0, \cdots, d_{q-p-r})) \\ &\quad = [c^r\sigma(d_{q-p-r}, \cdots, d_{q-p}) \circ d^p\sigma(d_{q-p}, \cdots, d_q) \circ b]\sigma(d_0, \cdots, d_{p-q-r}). \end{aligned}$$

On the other hand,

$$(c^r \smile d^p) \frown b\sigma = [(c^r \smile d^p)(\sigma(d_{q-r-p}, \cdots, d_q)) \circ b]\sigma(d_0, \cdots, d_{q-r-p}).$$

Here,

$$\begin{aligned} &(c^r \smile d^p)(d_{q-r-p}, \cdots, d_q) \\ &\quad = c^r(\sigma(d_{q-r-p}, \cdots, d_q)(d_0, \cdots, d_r) \circ d^p(\sigma(d_{q-r-p}, \cdots, d_q)(d_r, \cdots, d_{r+p}) \\ &\quad = c^r(\sigma(d_{q-r-p}, \cdots, d_{q-p})) \circ d_p(\sigma(d_{q-p}, \cdots, d_q). \end{aligned}$$

This completes the proof of the theorem.

21.9. Corollary: *If* $M = N$, M *is an associative ring and* P *is an* M*-module, then* $H_*(X, P)$ *is an* $H^*(X, M)$*-module.*

21.10. THEOREM: *Let c^0 be the cocycle whose value is 1 on each 0-simplex. Then c^0 is a unit element in $H^*(X, M)$ and $H^*(X, P)$ is a unitary module.*

Proof:

$$c^0 \frown b\sigma = [c^0(\sigma(d_q)) \circ b]\sigma(d_0, \cdots, d_q) = b\sigma.$$

21.11. Corollary: *If $M \times N \longrightarrow N$ is bilinear, M is an associative ring with a unit element, and N is a unitary M-module; then $H_*(X, N)$ is a unitary $H^*(X, M)$-module.*

21.12. THEOREM: *If $\alpha \in H_q(X)$, $\beta \in H^p(Y)$, and f_* and f^* are the homomorphisms of the homology and cohomology groups induced by a continuous function $f: X \longrightarrow Y$, then*

$$f_*(f^*\beta \frown \alpha) = \beta \frown f_*\alpha.$$

Proof: Let c^p be a cocycle in β, let σ be a p-simplex in X, and let $b \in N$. Then

$$\begin{aligned} f^\# c^p \frown b\sigma &= [f^\# c^p(\sigma(d_{q-p}, \cdots, d_q) \circ b]\sigma(d_0, \cdots, d_{q-p}) \\ &= [c^p f\sigma(d_{q-p}, \cdots, d_q) \circ b]\sigma(d_0, \cdots, d_{q-p}). \end{aligned}$$

Then

$$\begin{aligned} f_\#(f^\# c^p \frown d_q) &= [c^p f\sigma(d_{q-p}, \cdots, d_q) \circ b] f\sigma(d_0, \cdots, d_{q-p}) \\ &= c^p \frown f_* b\sigma, \end{aligned}$$

and this establishes the theorem.

Let $M = N = R$ be an associative ring. Let c^p be a cochain in $S^p(X)$ and let d_p be a chain in $S_p(X)$. Then $c^p \frown d_p \in S_0(X)$. In the module $S_0(X)$, the augmented boundary operator $\tilde{\partial}_0$ associates the empty simplex with each 0-simplex and associates the product of the coefficient sum and the empty simplex with each 0-chain. If $d_p = b\sigma$, then

$$c^p \frown d_p = (c^p\sigma \circ b)(\sigma(d_0)) \text{ and } \tilde{\partial}_0(c^p \frown b\sigma) = c^p\sigma \circ b = c^p(b\sigma).$$

By linearity, it follows for arbitrary d_p that

$$\tilde{\partial}_0(c^p \frown d_p) = c^p(d_p).$$

Therefore, *the question of whether* $c^p \frown d_p$ *is a cycle in the augmented theory can be settled by the value of* $c^p(d_p)$.

As an exercise, the reader can use the cap product to define the following products for space pairs (X, A):

$$H^p(X, A; M) \times H_q(X, A; N) \longrightarrow H_{q-p}(X, P)$$
$$H^p(X, M) \quad \times H_q(X, A; N) \longrightarrow H_{q-p}(X, A; P)$$

Part of the problem here is that $\sigma(d_0, \cdots, d_{q-p})$ could have a carrier in A even when $\sigma(d_0, \cdots, p_q)$ does not.

22

The Anticommutativity of the Cup Product

This chapter begins with a consideration of affine simplices.

Let $(a_0, \cdots, a_q)$ be an affine simplex $\Delta_q \longrightarrow E$; thus $(a_0, \cdots, a_q) \in A(\Delta_q, E)$.

Definitions: The homomorphism $\rho_q \colon A(\Delta_q, E) \longrightarrow A(\Delta_{q+1}, E)$ is defined by

$$\rho_q(a_0, \cdots, a_q) = \epsilon_q(a_q, \cdots, a_0),$$

where $\epsilon_q = (-1)^{q(q+1)/2}$.

The homomorphism $D_q \colon A(\Delta_q, E) \longrightarrow A(\Delta_q, E)$ is defined recursively by

$$D_{-1} = 0,\ D_0 = 0, \text{ and}$$
$$D_q(\sigma) = a_0(\sigma - \rho_q(\sigma) - D_{q-1}(\partial_q\sigma)), \text{ for } q \geq 1.$$

Here the statement $D_{-1} = 0$ is a formality. When $q = 1$,

$$D_1(\sigma) = (a_0, a_0, a_1) - (a_0, a_1, a_0).$$

22.1. Lemma: *If* $q \geq 1$, *then* $\partial_q\rho_q + \rho_{q-1}\partial_q$.
Proof:

$$\begin{aligned}\partial_q\rho_q(a_0, \cdots, a_q) &= \epsilon_q\partial_q(a_q, \cdots, a_0)\\ &= \sum_{i=1}^{q} \epsilon_q(-1)^{q-i}(a_q, \cdots, \hat{a}_i, \cdots, a_0),\end{aligned}$$

where $\hat{a}_i$ is located at the $(q - i)$th position. The lemma follows from the fact that $\epsilon_q(-1)^q = \epsilon_{q-1}$ and

$$\epsilon_{q-1}(a_q, \cdots, \hat{a}_i, \cdots, a_0) = \rho_{q-1}(a_0, \cdots, \hat{a}_i, \cdots, a_q).$$

22.2. Lemma: $\partial_{q+1}D_q + D_{q-1}\partial_q = \text{Id} - \rho_q$.

Proof: Since $D_{-1} = 0$ and $D_0 = 0$, the lemma is true when $q = 0$. Suppose that 22.2 has been proved for $q - 1$; in other words, that

$$\partial_q D_{q-1} + D_{q-2}\partial_{q-1} = \text{Id} - \rho_{q-1}.$$

A multiplication on the right by ∂_q yields

$$\partial_q D_{q-1}\partial_q = \partial_q - \rho_{q-1}\partial_q = \partial_q - \partial_q\rho_q.$$

By definition of D_q,

$$\partial_{q+1}D_q\sigma = \partial_{q+1}a_0(\sigma - \rho_q(\sigma) - D_{q-1}(\partial_q\sigma)).$$

Since $\partial_{q+1}a_0 + a_0\partial_q = \text{Id}$, it follows that

$$\begin{aligned}\partial_{q+1}D_q\sigma &= (\sigma - \rho_q(\sigma) - D_{q-1}(\partial_q\sigma)) - a_0(\partial_q\sigma - \partial_q\rho_q\sigma\\ &\qquad - (\partial_q - \partial_q\rho_q)(\sigma))\\ &= \sigma - \rho_q\sigma - D_{q-1}\partial_q\sigma.\end{aligned}$$

This establishes 22.2.

Notice that the recursion formula 22.2 shows that $D_q\sigma$ is a linear combination of $(q + 1)$-simplices with coefficients that are ± 1 and that the vertices of these simplices are among $a_0, \cdots, a_q$. Now let the space E be restricted to the simplex Δ_q, so that $\rho_q(\Delta_q^\dagger) \in A(\Delta_q, \Delta_q)$ and $D_q(\Delta_q^\dagger) \in A(\Delta_{q+1}, \Delta_q)$.

22.3. Lemma: $\rho_q\sigma = \sigma\rho_q(\Delta_q^\dagger)$ *and* $D_q\sigma = \sigma D_q(\Delta_q^\dagger)$.

Proof: The first part of 22.3 follows immediately from the definitions of the symbols involved. For the second part, let $\nu_0, \cdots, \nu_{q+1}$ be integers between 0 and q. Then

$$(a_0, \cdots, a_q)(d_{\nu_0}, \cdots, d_{\nu_q}) = (a_{\nu_0}, \cdots, a_{\nu_q})$$

and

$$(a_0, \cdots, a_q)(d_{\nu_0}, \cdots, d_{\nu_{q+1}}) = (a_{\nu_0}, \cdots, a_{\nu_{q+1}}).$$

Hence,

$$\begin{aligned} D_q(\Delta_q^\dagger)\partial_{q+1}(\Delta_{q+1}^\dagger) + \partial_q(\Delta_q^\dagger)D_{q-1}(\Delta_{q-1}^\dagger) &= \partial_{q+1}D_q(\Delta_q^\dagger) + D_{q-1}\partial_q(\Delta_q^\dagger) \\ &= (\mathrm{Id} - \rho_q)(\Delta_q^\dagger) \\ &= \Delta_q^\dagger - \rho_q(\Delta_q^\dagger). \end{aligned}$$

Now the considerations shift from affine simplices to space pairs and singular homology. Let (X, A) be a space pair, and let $S_q(X, A)$ be formed with coefficients in the ring Z of integers. The reader will remember that $S_q(X, A)$ is the factor group $S_q(X)/S_q(A)$.

Definition: If $\sigma \in S_q(X)$, let

$$\rho_q\sigma = \begin{cases} \mathrm{Id} & \text{when } q \leq 0 \\ \sigma\rho_q(\Delta_q^\dagger) & \text{when } q \geq 0, \end{cases}$$

and

$$D_q\sigma = \begin{cases} 0 & \text{when } q \leq 0 \\ \sigma D_q(\Delta_q^\dagger) & \text{when } q \geq 0. \end{cases}$$

22.4. Lemma: *Equation* 22.2 *holds under the extended definitions of* ρ_q *and* D_q*; that is,*

$$\partial_{q+1}D_q + D_{q-1}\partial_q = \text{Id} - \rho_q.$$

Proof: If $q \leq 0$,

$$\partial_{q+1}D_q\sigma + D_{q-1}\partial_q\sigma = 0 = (\text{Id} - \rho_q)\sigma.$$

If $q \geq 0$, then

$$\begin{aligned}\partial_{q+1}D_q\sigma + D_{q-1}\partial_q\sigma &= \sigma\{(D_q(\Delta_q^\dagger)\partial_{q+1}(\Delta_{q+1}^\dagger) + \partial_q(\Delta_q^\dagger)D_{q-1}(\Delta_{q-1}^\dagger)\} \\ &= \sigma(\Delta_q^\dagger - \rho_q(\Delta_q^\dagger)) \\ &= \sigma - \rho_q\sigma.\end{aligned}$$

22.5. Corollary: *The homomorphisms ρ_q and D_q induce homomorphisms (of the same name) on the factor group $S_q(X)/S_q(A)$ which also satisfy* 22.2.

Let M be an R-module. Then the cochain complex $S_q(X, A; M)$ was defined to be Hom $(S_q(X, A), M)$.

Definition: The homomorphisms ρ_q and D_q induce homomorphisms (with reversed direction)

$$\rho_q^\#: S^q(X, A; M) \longrightarrow S^q(X, A; M)$$

and

$$D_q^\#: S^{q+1}(X, A; M) \longrightarrow S^q(X, A; M)$$

by means of

$$(\rho_q^\# f^q)c_q = f^q(\rho_q c_q)$$

and

$$(D_{q+1}^\# f^{q+1})c_q = f^{q+1}(D_q c_q),$$

where $c_q \in S_q(X, A)$ and $f^{q+1} \in S^{q+1}(X, A; M)$.

22.6. Lemma: $\delta D_q^\# + D_{q+1}^\#\delta = \text{Id} - \rho_q$.

Proof: Consider the image of the cochain f^{q+1} under the left-hand side of 22.6. On the chain c_{q+1}, this has the value

$$\begin{aligned}((\delta D_q^{\#} + D_{q+1}^{\#}\delta)f^q)c_q) &= (\delta(D_q^{\#}f^q))(c_q) + (D_{q+1}^{\#}(\delta f^q))(c_q)\\ &= D_q^{\#}f^q(\partial c_q) + \delta f^q(D_q c_q)\\ &= f^q(D_{q-1}\partial c_q + \partial D_q c_q)\\ &= f^q((\mathrm{Id} - \rho_q)c_q)\\ &= ((\mathrm{Id} - \rho_q^{\#})f^q)(c_q).\end{aligned}$$

Since $\rho_q^{\#}$ is chain homotopic to the identity by 22.6, it follows at once that $\rho_q^* = \mathrm{Id}$. This means that if f^q is a representative of the cohomology class α, then $\rho_q^* f^q$ is also a representative. Furthermore,

22.7. $$\begin{aligned}(\rho_q^{\#}f^q)(\sigma) &= f^q(\rho_q(\sigma))\\ &= \textstyle\sum_q f^q(\sigma\rho_q(\Delta_q^\dagger)) = \sum_q f^q(\sigma(d_q, \cdots, d_0)).\end{aligned}$$

The tools needed for the investigation of the anticommutativity of the cup product are now at hand.

22.8. THEOREM: *If $M = N$ and multiplication M is commutative, or if a product $n \circ m$ where $n \in N$ and $m \in M$ is suitably defined, then*

$$\alpha^p \smile \beta^q = (-1)^{pq}(\beta^q \smile \alpha^p).$$

Proof: Let f^p and g^q be representatives of the cohomology classes α and β respectively, where

$$\alpha \in H^p(X, A; M) \text{ and } \beta \in H^q(X, A; N).$$

Then $\rho_p^{\#}f^p$ and $\rho_q^{\#}g^q$ also serve as representatives, and the cohomology class $\alpha \smile \beta$ is represented by

$$(\rho_p^{\#}f^p) \smile (\rho_q^{\#}g^q).$$

The value of this cochain on the simplex (chain) σ is

$$\begin{aligned}(\rho^{\#}f^p \smile \rho^{\#}g^q)(\sigma) &= \rho^{\#}f^p(\sigma(d_0, \cdots, d_p)) \circ \rho^{\#}g^q(\sigma(d_p, \cdots, d_{p+q}))\\ &= \epsilon_p\epsilon_q(f^p(\sigma(d_0, \cdots, d_p)(d_p, \cdots, d_0))) \circ\\ &\qquad\qquad g^q(\sigma(d_p, \cdots, d_{p+q})(d_q, \cdots, d_0)))\\ &= \epsilon_p\epsilon_q(f^p\sigma(d_p, \cdots, d_0)) \circ (g^q\sigma(d_{p+q}, \cdots, d_p)).\end{aligned}$$

The homomorphism $\rho_{p+q}^{\#}$ can be applied to this result without changing the cohomology class of $\rho^{\#}f^p \smile \rho^{\#}g^q$. If this is carried out, the sign will be changed by a factor of ϵ_{p+q} and the simplex σ will be replaced by $\sigma(d_{p+q}, \cdots, d_0)$. Since

$$\epsilon_{p+q} = (-1)^{pq}\epsilon_p\epsilon_q$$

and

$$(d_{p+q}, \cdots, d_0)(d_p, \cdots, d_0) = (d_q, \cdots, d_{p+q})$$
$$(d_{p+q}, \cdots, d_0)(d_{p+q}, \cdots, d_p) = (d_0, \cdots, d_q),$$

it follows that the value of a representative of $\alpha \smile \beta$ on σ is

$$(-1)^{pq}((f^p\sigma(d_q, \cdots, d_{p+q})) \circ (g^q\sigma(d_0, \cdots, d_q)).$$

The multiplication of M and N was of the form $M \times N \longrightarrow P$. This permits the definition of a product $N \times M \longrightarrow P$ by means of

$$n \circ m = m \circ n.$$

(If $N = M$, this definition is unambiguous only if multiplication is commutative.) The new multiplication permits the definition of $\beta \smile \alpha$ and the computations just carried out prove Theorem 22.8 when $q \geq 0$ and $p \geq 0$. The theorem is trivial when $q < 0$ or $p < 0$.

The reader who is familiar with the definition of a grassmannian algebra will see that the graded cohomology ring with coefficients in a commutative ring is such an algebra.

Index

Index

Additive functor 109
Adjoin 146
Admissible
—functions 130
—pairs 130
—triples 130
Affine
—map 14
—simplex 20
—space 10
Antipodal function 79
Associative law
—for cup product 207
—for tensor product 87
Attaching of n-cells 158
Augmented complex 29
Axioms
—for a category 107
—of Eilenberg and Steenrod 129 ff.

Barycenter 57
Barycentric coordinates 12
Basis 18
Betti number 169
Bilinear 19, 86
Boundary
—of affine simplex 22
—of prism 41
—of singular chain 28
—operator 4
Brouwer 152

Cap product 209
Cauchy integral theorem 126
Carrier 28
Category 107
Linear— 108
Chain 27
—complex 4
—homomorphism 4
—homotopy 43
Coefficient module 131
Cohomology 120
Cohomology group
Axioms for— 130 ff.
Singular— 121
—of a point 122
—of CP^n 177 ff.
—of S^n 122
—of P^n 180
Cokernel 131
Commutative diagram 4
Commutative law
—for tensor products 88
—for cup products 220
Complex
Augmented— 27
Chain— 4
CW— 164
Homomorphism of— 4
Simplicial— 194
Spherical— 164
Complex projective n-space 177
Component, path— 71 ff.
Connected, n-tuply— 81
Contravariant 98
—functor 109
Convex 12
—hull 13
Coordinate vector 11
Covariant 98

—functor 108
Cup product 202
Cycle 4

Deformation retract 51
Strong— 51
Degree 81
Diagram 3
Commutative— 4
Nine— 92
Diameter 61
Dimension
Invariance of— 157
—axiom 131
—of an affine space 12
Direct
—decomposition theorem 71, 120
—product 17
—sum 17, 70
Divisible module 95
Domain, invariance of— 156

Eilenberg 129
Epimorphism 2
Equator 75
Equivalence
Homotopy— 49
—in a category 108
Euler characteristic 170
Exact
—sequence 3
—triad 141
Excision 74
—axiom 131
—theorem 67, 117

Five lemma 55
Free module 18
Fox, R. 154
Function
Affine— 14
Bilinear— 19
Homotopy of a— 42
Multilinear— 85
—of a space tuple 32
Functor
Additive— 109
Contravariant— 109
Covariant— 109

Genus 146
Generating system 18
Graded module 205
Graph 53
Homology groups of— 80
Application of Mayer-Vietoris theorem to— 145

Hemisphere 75
Hexagon theorem 137
Homologous to zero 125
Homology
—ring 207
—sequence 6, 130
—sequence of a triad 140
—sequence of a triple 35
Homology group 4
Augmented— 29
Reduced— 29
Singular— 29
—of a chain complex 4
—of a pair 31
—of a point 73, 120, 131
—of CP^n 177
—of P^n 178
—of S^n 77
Homomorphism 4
Chain— 4
Homotopy 42
Chain— 43
—axiom 131
—equivalence 49
—inverse 49
—of space pairs 49
—theorem 47, 117
Hull, convex— 13

Independent points 12
Injective
—module 163
—representation 112
Inverse
Homotopy— 49
—of a morphism 108
Isomorphism 2

Jordan-Brouwer separation theorem 152

Kernel 1
Klein bottle 197
Knot 155
Artin-Fox— 154

Lens space 189
homology of— 190
Line 11
Linear category 108

Manifold 173
Mayer-Vietoris
—sequence 143
—theorem 153
—theorem, relative case 147
Module
Basis for— 18
Coefficient— 131
Divisible— 95
Free— 18
Generating system for— 118
Graded— 205
Homorphism of graded— 205
Injective— 163
Projective— 163
Quotient— 106
R— 1
(*R*,*S*)— 101
Torsion— 96, 103
Torsion free— 103
Monomorphism 2
Morphism 107
Multilinear 85

Nine diagram 92

Path-component 71 ff.
Prism construction 39
Product
Cap— 209
Cup— 202
Direct— 17
Tensor— 86
—of functions 91
Projective
—module 163
—plane 172
—representation 112
Projective spaces 175 ff.
Euler characteristics of— 180
Homology groups of— 180
Proper triad 141

Quotient module 106

Real projective *n*-space 178
Region 155
Representation
Injective— 112
Projective— 112
Retract 50
Deformation— 51

Sequence 2
Exact— 3, 5
Mayer-Vietoris— 143
Simplex 14
Affine— 20
Empty— 20
Singular— 27
Sphere 75, Chapter 18
Homology groups of— 77, 144
Separation on— 150 ff.
Steenrod 129
Subdivision 57
Sum, direct— 17

Surface 173
    Classification of— 195 ff.
    —of genus g 146
Tensor product 86
Torsion 103
Torus 172
Triad 140
    Exact— 141
    Proper— 141
Triangulable space pair 194
Triple, homology sequence of 35

Vector field 79
Vertex 14
Vietoris 143

Index of Symbols

$\frown$ 209
$\smile$ 209

(.) 20
f_* 4
$f^\#$ 33
$f(X, A_1, \ldots, A_r) \rightarrow (Y, B_1, \ldots B_r)$ 31

$A_0(X, Y)$ 19
$A(X, Y)$ 19
$\mathscr{C}(A)$ 141
$C(X, Y)$ 19
CP^n 175
D_q 65
F_g 173
$H^*(X, M)$ 213
$H_q(X, A)$ 30
$H^q(X, A)$ 12
Hom 2, 98
$L(n, m)$ 189
P^n 175
Sd 58
S^n 75
$S_q(X)$ 27
∂ 4, 23
∂_* 5
χ 179
ψ_q 39
Δ_q 20
$\Delta_q^\dagger$ 24
Φ_h 173